Introduction to Internal Combustion Engines

Other Macmillan titles of related interest

E. M. Goodger, *Combustion Calculations: Theory, worked examples and problems*

E. M. Goodger, *Principles of Engineering Thermodynamics*, second edition

E. M. Goodger and R. A. Vere, *Aviation Fuels Technology*

Richard T. C. Harman, *Gas Turbine Engineering: Applications, cycles and characteristics*

N. Watson and M. S. Janota, *Turbocharging the Internal Combustion Engine*

Introduction to
Internal Combustion Engines

Richard Stone
Brunel University
Uxbridge, Middlesex

MACMILLAN

First published 1985

Published by
Higher and Further Education Division
MACMILLAN PUBLISHERS LTD
Houndmills, Basingstoke, Hampshire RG21 2XS
and London
Companies and representatives
throughout the world

Printed in Hong Kong

British Library Cataloguing in Publication Data

Stone, Richard
    Introduction to internal combustion engines.
    1. Internal combustion engine
    I. Title
    621.43   TJ785

    ISBN 0-333-37593-9
    ISBN 0-333-37594-7 Pbk

Contents

CONTENTS ix

Preface

This book aims to provide for students and engineers the background that is presupposed in many articles, papers and advanced texts. Since the book is primarily aimed at students, it has sometimes been necessary to give only outline or simplified explanations. However, numerous references have been made to sources of further information.

Internal combustion engines form part of most thermodynamics courses at Polytechnics and Universities. This book should be useful to students who are following specialist options in internal combustion engines, and also to students at earlier stages in their courses — especially with regard to laboratory work.

Practising engineers should also find the book useful when they need an overview of the subject, or when they are working on particular aspects of internal combustion engines that are new to them.

The subject of internal combustion engines draws on many areas of engineering: thermodynamics and combustion, fluid mechanics and heat transfer mechanics, stress analysis, materials science, electronics and computing. These disparate areas are drawn together in the first eight chapters, so that these chapters are best read in sequence. However, internal combustion engines are not just subject to thermodynamic or engineering considerations — the commercial (marketing, sales etc.) and economic aspects are also important, and these are discussed as they arise.

Chapter 1 provides an introduction, with definitions of engine types and operating principles. The essential thermodynamics is provided in chapter 2, while chapter 3 provides the background in combustion and fuel chemistry. The differing needs of spark ignition engines and compression ignition engines are discussed in chapters 4 and 5 respectively.

Chapter 6 describes how the induction and exhaust processes are controlled, and this leads to chapter 7, where turbochargers are discussed. The remaining chapters can be read in parallel with the earlier chapters. Some of the mechanical and materials aspects are discussed in chapter 8, while chapter 9 covers some experimental techniques. Finally, chapter 10 provides three case studies that should remain topical for some time.

This book is the product of information gained from numerous sources, and it would be invidious to acknowledge individuals. However, I should like to express my gratitude to Dr Neil Watson of Imperial College, London, and to Dr Neil Richardson of Jesus College, Oxford, for reading and commenting on the drafts. I must also express my thanks to the typists, in particular to Mrs Gill Oxley who typed most of the material.

In conclusion, I would welcome any criticism and suggestions concerning either the detail of the book or the overall concept.

Autumn 1984   RICHARD STONE

Acknowledgements

The author and publisher wish to thank the following, who have kindly given permission for the use of copyright material.

Dr W. J. D. Annand (University of Manchester) for seven figures from W. J. D. Annand and G. E. Roe, *Gas Flow in the Internal Combustion Engine*, published by Foulis, Yeovil, 1974.

Atlantic Research Associates, Tunbridge Wells, and Martin H. Howarth for three figures from M. H. Howarth, *The Design of High Speed Diesel Engines*, published by Constable, London, 1966.

Blackie and Son Ltd, Glasgow, for three figures from H. R. Ricardo and J. G. G. Hempson, *The High Speed Internal Combustion Engine*, 1980.

Butterworths, Guildford, for five figures from K. Newton, W. Steeds and T. K. Garrett, *The Motor Vehicle*, 10th edn, 1983.

Ford of Europe, Inc., Brentwood, for eleven figures from Ford technical publications.

Froude Consine Ltd, Worcester, for two figures from *Publication No. 526/2*.

GKN Engine Parts Division, Maidenhead, for a figure from *Publication No. EPD 82100*.

Hutchinson Publishing Group Ltd, London, for a figure from A. Baker, *The Component Contribution* (co-sponsored by the AE Group), 1979.

Johnson Matthey Chemicals Ltd, Royston, for three figures from *Catalyst Systems for Exhaust Emission Control from Motor Vehicles*.

Longmans Group Ltd, Harlow, for four figures from G. F. C. Rogers and Y. R. Mayhew, *Engineering Thermodynamics*, 3rd end, 1980; two figures from H. Cohen, G. F. C. Rogers and H. I. H. Saravanamuttoo, *Gas Turbine Theory*, 2nd edn, 1972.

Lucas CAV Ltd, London, for eight figures from *CAV Publications 586, 728, 730, 773* and *C2127E*, and a press release photograph.

Lucas Electrical Ltd, Birmingham, for two figures from publications *PLT 6339* (Electronic Fuel Injection) and *PLT 6176* (Ignition).

MIT Press, Cambridge, Massachusetts, and Professor C. F. Taylor for a figure from C. F. Taylor, *The Internal Combustion Engine in Theory and Practice*, 1966/8.

Oxford University Press for two figures from Singer's *History of Technology*.

Patrick Stephens Ltd, Cambridge, for three figures from A. Allard, *Turbocharging and Supercharging*, 1982.

Pergamon Press Ltd, Oxford, for a figure from R. S. Benson and N. D. Whitehouse, *Internal Combustion Engines*, 1979; three figures from H. Daneshyar, *One-Dimensional Compressible Flow*, 1976.

Plenum Press, New York, and Professor John B. Heywood (MIT) for a figure from J. B. Heywood, *Combustion Modelling in Reciprocating Engines*, 1980.

Purnell Books, Bristol, for a figure from A. F. Evans, *The History of the Oil Engine*.

Society of Automotive Engineers, Warrendale, Pennsylvania, for ten figures and copy from *SAE 790291, SAE 820167, SAE 821578* and *Prepr. No. 61A* (1968).

Sulzer Brothers Ltd, Winterthur, Switzerland, for three figures from the *Sulzer Technical Review*, Nos 1 and 3 (1982).

Material is acknowledged individually throughout the text of the book.

Every effort has been made to trace all copyright holders but, if any have been inadvertently overlooked, the publisher will be pleased to make the necessary arrangements at the first opportunity.

Notation

a	sonic velocity (m/s)
abdc	after bottom dead centre
atdc	after top dead centre
A	piston area (m²)
A_c	curtain area for poppet valve (m²)
A_e	effective flow area (m²)
A_f	flame front area (m²)
A_o	orifice area (m²)
A/F	air/fuel ratio
bbdc	before bottom dead centre
bdc	bottom dead centre
bmep	brake mean effective pressure (N/m²)
btdc	before top dead centre
BHP	brake horse power
c_p	specific heat capacity at constant pressure (kJ/kg K)
c_v	specific heat capacity at constant volume (kJ/kg K)
C_D	discharge coefficient
C_o	orifice discharge coefficient
C_p	molar heat capacity at constant pressure (kJ/kg K)
C_v	molar heat capacity at constant volume (kJ/kg K)
CI	compression ignition
CV	calorific value (kJ/kg)
dc	direct current
dohc	double overhead camshaft
D_v	valve diameter (m)
DI	direct injection (compression ignition engine)
E	absolute internal energy (kJ)
EGR	exhaust gas recirculation
f	fraction of exhaust gas residuals
ff	turbulent flame factor
fwd	front wheel drive
g	gravitational acceleration (m/s²)

G	Gibbs function (kJ)
h	specific enthalpy (kJ/kg); manometer height (m)
h_d	mean height of indicator diagram (m)
H	enthalpy (kJ)
imep	indicated mean effective pressure (N/m^2)
I	current (A)
IDI	indirect injection (compression ignition engine)
jv	just visible (exhaust smoke)
k	constant
K	equilibrium constant
l	length, connecting-rod length (m)
l_b	effective dynamometer lever arm length (m)
l_d	indicator diagram length (m)
L	stroke length (m); inductance (H)
L_D	duct length (m)
L_v	valve lift (m)
LDA	laser Doppler anemometer
LDV	laser Doppler velocimeter
m	mass (kg)
$\dot{m}_a$	air mass flow rate (kg/s)
$\dot{m}_f$	fuel mass flow rate (kg/s)
m_r	reciprocating mass (kg)
M	mutual inductance (H)
MBT	minimum (ignition) advance for best torque (degrees)
n	number of moles/cylinders
N^*	rev./s for 2-stroke, rev./2s for 4-stroke engines
N'	total number of firing strokes/s ($\equiv n.N^*$)
Nu	Nusselt number (dimensionless heat transfer coefficient)
ohc	overhead camshaft
ohv	overhead valve
p	pressure (N/m^2)
p'	partial pressure (N/m^2)
$\overline{p}_b$	brake mean effective pressure (N/m^2)
Pr	Prandtl number (ratio of momentum and thermal diffusivities)
Q	heat flow (kJ)
r	crankshaft throw ($\equiv \frac{1}{2}$ engine stroke) (m)
r_v	volumetric compression ratio
R	specific gas constant (kJ/kg K)
R_0	molar (or universal) gas constant (kJ/kmol K)
Re	Reynolds number
s	specific entropy (kJ/kg K)
sfc	specific fuel consumption (MJ/kg)
SI	spark ignition
t	time (s)

tdc	top dead centre
T	absolute temperature (K); torque (N m)
T_0	absolute temperature of the environment (K)
U_l	laminar flame front velocity (m/s)
U_t	turbulent flame front velocity (m/s)
v	velocity (m/s)
v_p	mean piston velocity (m/s)
V	volume (m^3)
$\dot{V}_a$	volumetric flow rate of air (m^3/s)
V_s	engine swept volume (m^3)
wmmp	weakest mixture for maximum power
W	work (kJ)
$\dot{W}$	power (kW)
W_b	brake work (kJ)
W_c	compressor work (kJ)
W_f	friction work (kJ)
W_i	indicated work (kJ)
W_{REV}	work output from a thermodynamically reversible process (kJ)
W_t	turbine work (kJ)
WOT	wide open throttle
x	a length (m); mass fraction
α	cut off (or load) ratio
γ	ratio of gas heat capacities, c_p/c_v or C_p/C_v
ΔH_0	enthalpy of reaction (combustion) $\equiv$ −CV
Δp	pressure difference (N/m^2)
$\Delta\theta_b$	combustion duration (crank angle, degrees)
ϵ	heat exchanger effectiveness = (actual heat transfer)/(max. possible heat transfer)
η	efficiency
η_b	brake thermal efficiency $\equiv \eta_o$
η_c	isentropic compressor efficiency
η_{Diesel}	ideal air standard Diesel cycle efficiency
η_{FA}	fuel–air cycle efficiency
η_i	indicated (arbitrary overall) efficiency
η_m	mechanical efficiency
η_o	arbitrary overall efficiency
η_{Otto}	ideal air standard Otto cycle efficiency
η_R	rational efficiency, W/W_{REV}
η_t	isentropic turbine efficiency
η_v	volumetric efficiency
θ	crank angle (degrees)
θ_0	crank angle at the start of combustion (degrees)
μ	dynamic viscosity (N s/m)
ρ	density (kg/m^3)

ρ_u density of the unburnt gas (kg/m^3)

ϕ equivalence ratio = (stoichiometric air/fuel ratio)/(actual air/fuel ratio)
(*note that* sometimes the reciprocal definition is used in other publications)

ω specific humidity (kg water/kg dry air); angular velocity (rad/s)

1 Introduction

1.1 Fundamental operating principles

The reciprocating internal combustion engine must be by far the most common
form of engine or prime mover. As with most engines, the usual aim is to achieve
a high work output with a high efficiency; the means to these ends are developed
throughout this book. The term 'internal combustion engine' should also include
open circuit gas turbine plant where fuel is burnt in a combustion chamber. However, it is normal practice to omit the prefix 'reciprocating'; none the less this is
the key principle that applies to both engines of different types and those
utilising different operating principles. The divisions between engine types and
between operating principles can be explained more clearly if stratified charge
and Wankel-type engines are ignored initially; hence these are not discussed until
section 1.4.

The two main types of internal combustion engine are: spark ignition (SI)
engines, where the fuel is ignited by a spark; and compression ignition (CI)
engines, where the rise in temperature and pressure during compression is
sufficient to cause spontaneous ignition of the fuel. The spark ignition engine
is also referred to as the petrol, gasoline or gas engine from its typical fuels, and
the Otto engine, after the inventor. The compression ignition engine is also
referred to as the Diesel or oil engine; the fuel is also named after the inventor.

During each crankshaft revolution there are two strokes of the piston, and
both types of engine can be designed to operate in either four strokes or two
strokes of the piston. The four-stroke operating cycle can be explained by
reference to figure 1.1.

(1) The induction stroke. The inlet valve is open, and the piston travels down
     the cylinder, drawing in a charge of air. In the case of a spark ignition
     engine the fuel is usually pre-mixed with the air.
(2) The compression stroke. Both valves are closed, and the piston travels up the
     cylinder. As the piston approaches top dead centre (tdc), ignition occurs. In
     the case of compression ignition engines, the fuel is injected towards the
     end of the compression stroke.

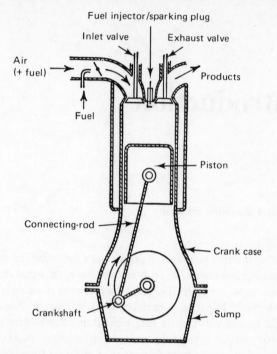

Figure 1.1   A four-stroke engine (reproduced with permission from Rogers and
            Mayhew (1980a))

(3) The expansion, power or working stroke. Combustion propagates through-
     out the charge, raising the pressure and temperature, and forcing the piston
     down. At the end of the power stroke the exhaust valve opens, and the
     irreversible expansion of the exhaust gases is termed 'blow-down'.
(4) The exhaust stroke. The exhaust valve remains open, and as the piston
     travels up the cylinder the remaining gases are expelled. At the end of the
     exhaust stroke, when the exhaust valve closes some exhaust gas residuals
     will be left; these will dilute the next charge.

The four-stroke cycle is sometimes summarised as 'suck, squeeze, bang and blow'.
Since the cycle is completed only once every two revolutions the valve gear (and
fuel injection equipment) have to be driven by mechanisms operating at half
engine speed. Some of the power from the expansion stroke is stored in a flywheel, to provide the energy for the other three strokes.

The two-stroke cycle eliminates the separate induction and exhaust strokes;
and the operation can be explained with reference to figure 1.2.

(1) The compression stroke (figure 1.2a). The piston travels up the cylinder, so
     compressing the trapped charge. If the fuel is not pre-mixed, the fuel is
     injected towards the end of the compression stroke; ignition should again

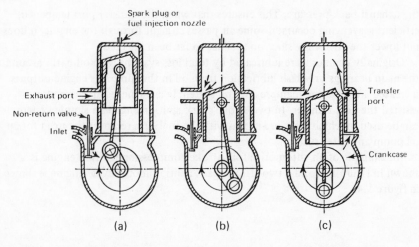

Figure 1.2   A two-stroke engine (reproduced with permission from Rogers and
Mayhew (1980a))

occur before top dead centre. Simultaneously, the underside of the piston
is drawing in a charge through a spring-loaded non-return inlet valve.

(2)   The power stroke. The burning mixture raises the temperature and pressure
in the cylinder, and forces the piston down. The downward motion of the
piston also compresses the charge in the crankcase. As the piston approaches
the end of its stroke the exhaust port is uncovered (figure 1.2b) and blow-
down occurs. When the piston is at bottom dead centre (figure 1.2c) the
transfer port is also uncovered, and the compressed charge in the crankcase
expands into the cylinder. Some of the remaining exhaust gases are displaced by the fresh charge; because of the flow mechanism this is called
'loop scavenging'. As the piston travels up the cylinder, first the transfer
port is closed by the piston, and then the exhaust port is closed.

For a given size engine operating at a particular speed, the two-stroke engine
will be more powerful than a four-stroke engine since the two-stroke engine has
twice as many power strokes per unit time. Unfortunately the efficiency of a
two-stroke engine is likely to be lower than that of a four-stroke engine. The
problem with two-stroke engines is ensuring that the induction and exhaust
processes occur efficiently, without suffering charge dilution by the exhaust gas
residuals. The spark ignition engine is particularly poor, since at part throttle
operation the crankcase pressure can be less than atmospheric pressure. This
leads to poor scavenging of the exhaust gases, and a rich air/fuel mixture
becomes necessary for all conditions, with an ensuing low efficiency (see chapter 4, section 4.1).

These problems can be overcome in two-stroke compression ignition engines
by supercharging, so that the air pressure at inlet to the crankcase is greater than

the exhaust back-pressure. This ensures that when the transfer port is opened, efficient scavenging occurs; if some air passes straight through the engine, it does not lower the efficiency since no fuel has so far been injected.

Originally engines were lubricated by total loss systems with oil baths around the main bearings or splash lubrication from oil in the sump. As engine outputs increased a circulating high-pressure oil system became necessary; this also assisted the heat transfer. In two-stroke spark ignition engines a simple system can be used in which oil is pre-mixed with the fuel; this removes the need for an oil pump and filter system.

An example of an automotive four-stroke compression ignition engine is shown in figure 1.3, and a two-stroke spark ignition motor cycle engine is shown in figure 1.4.

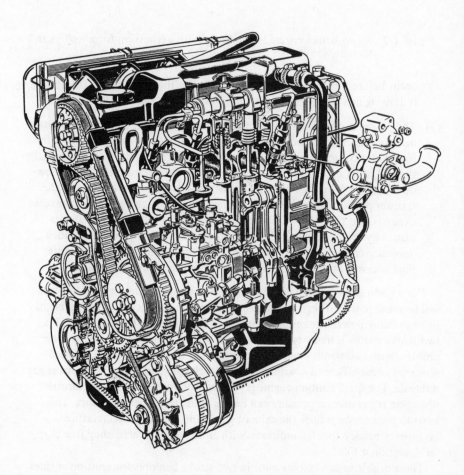

Figure 1.3   Ford 1.6 litre indirect injection Diesel (courtesy of Ford)

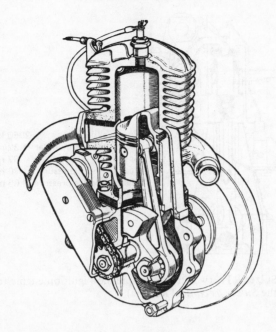

Figure 1.4 Two-stroke spark ignition engine

The size range of internal combustion engines is very large, especially for compression ignition engines. Two-stroke compression ignition engines vary from engines for models with swept volumes of about 1 cm^3 and a fraction of a kilowatt output, to large marine engines with a cylinder bore of about 1 m, up to 12 cylinders in-line, and outputs of up to 50 MW.

An example of a large two-stroke engine is the Sulzer RTA engine (see figure 1.5) described by Wolf (1982). The efficiency increases with size because the effects of clearances and cooling losses diminish. As size increases the operating speed reduces and this leads to more efficient combustion; also the specific power demand from the auxiliaries reduces. The efficiency of such an engine can exceed 50 per cent, and it is also capable of burning low-quality residual fuels. A further advantage of a low-speed marine diesel engine is that it can be coupled directly to the propeller shaft. To run at low speeds an engine needs a long stroke, yet a stroke/bore ratio greater than 2 leads to poor loop scavenging.

The Sulzer RTA engine has a stroke/bore ratio of about 3 and uses uniflow scavenging. This requires the additional complication of an exhaust valve. Figure 1.6 shows arrangements for loop, cross and uniflow scavenging. The four-stroke compression ignition engines have a smaller size range, from about 400 cm^3 per cylinder to 60 litres per cylinder with an output of 600 kW per cylinder at about 600 rpm with an efficiency of over 45 per cent.

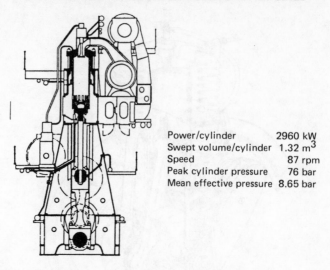

Power/cylinder	2960 kW
Swept volume/cylinder	1.32 m^3
Speed	87 rpm
Peak cylinder pressure	76 bar
Mean effective pressure	8.65 bar

Figure 1.5   Sulzer RTA two-stroke compression ignition engine (courtesy of
the *Sulzer Technical Review*)

The size range of two-stroke spark ignition engines is small, with the total
swept volumes rarely being greater than 1000 cm^3. The common applications
are in motor cycles and outboard motors, where the high output, simplicity and
low weight are more important than their poor fuel economy. Automotive four-
stroke spark ignition engines usually have cylinder volumes in the range 50–
500 cm^3, with the total swept volume rarely being greater than 5000 cm^3. Engine
outputs are typically 45 kW/litre, a value that can be increased seven-fold by
tuning and turbocharging.

The largest spark ignition engines are gas engines; these are usually converted
from large medium-speed compression ignition engines. High-output spark
ignition engines have been developed for racing and in particular for aero-engines.
A famous example of an aero-engine is the Rolls Royce Merlin V12 engine. This
engine had a swept volume of 27 litres and a maximum output of 1.48 MW at
3000 rpm; the specific power output was 1.89 kW/kg.

1.2 Early internal combustion engine development

As early as 1680 Huygens proposed to use gunpowder for providing motive
power. In 1688 Papin described the engine to the Royal Society of London, and
conducted further experiments. Surprising as it may seem, these engines did not
use the expansive force of the explosion directly to drive a piston down a

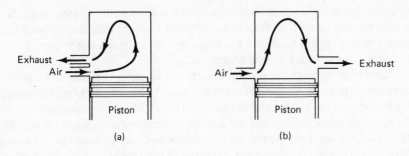

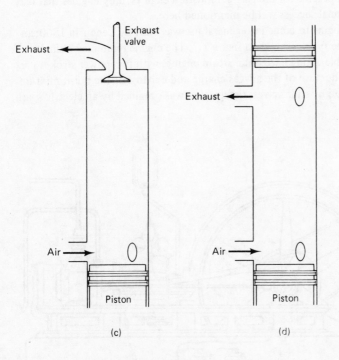

Figure 1.6   Two-stroke scavenging systems. (a) Loop scavenging; (b) cross
scavenging; (c) uniflow scavenging with exhaust valve; (d) uniflow
scavenging with opposed pistons

cylinder. Instead, the scheme was to explode a small quantity of gunpowder in a cylinder, and to use this effect to expel the air from the cylinder. On cooling, a partial vacuum would form, and this could be used to draw a piston down a cylinder – the so-called 'atmospheric' principle.

Papin soon found that it was much more satisfactory to admit steam and condense it in a cylinder. This concept was used by Newcomen who constructed his first atmospheric steam engine in 1712. The subsequent development of atmospheric steam engines, and the later high-pressure steam engines (in which the steam was also used expansively), overshadowed the development of internal combustion engines for almost two centuries. When internal combustion engines were ultimately produced, the technology was based heavily on that of steam engines.

Throughout the late 18th and early 19th century there were numerous proposals and patents for internal combustion engines; only engines that had some commercial success will be mentioned here.

The first engine to come into general use was built by Lenoir in 1860; an example of the type is shown in figure 1.7. The engine resembled a single-cylinder, double-acting horizontal steam engine, with two power strokes per revolution. Induction of the air/gas charge and exhaust of the burnt mixture were controlled by slide valves; the ignition was obtained by an electric spark.

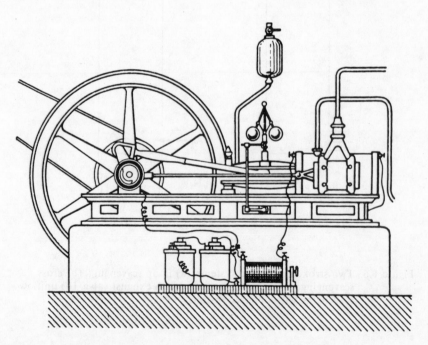

Figure 1.7   Lenoir gas-engine of 1860 (reproduced with permission from Singer, *History of Technology*, OUP, 1958)

Combustion occurred on both sides of the piston, but considering just one combustion chamber the sequence was as follows.

(1) In the first part of the stroke, gas and air were drawn in. At about half stroke, the slide valves closed and the mixture was ignited; the explosion then drove the piston to the bottom of the stroke.

(2) In the second stroke, the exhaust gases were expelled while combustion occurred on the other side of the piston.

The next significant step was the Otto and Langen atmospheric or free-piston engine of 1866; the fuel consumption was about half that of the Lenoir engine. The main features of the engine were a long vertical cylinder, a heavy piston and a racked piston rod (figure 1.8). The racked piston rod was engaged with a pinion connected to the output shaft by a ratchet. The ratchet was arranged to free-wheel on the upward stroke, but to engage on the downward stroke. Starting with the piston at the bottom of the stroke the operating sequence was as follows.

(1) During the first tenth or so of the stroke, a charge of gas and air was drawn into the cylinder. The charge was ignited by a flame transferred through a slide valve, and the piston was forced to the top of its stroke without delivering any work, the work being stored as potential energy in the heavy piston.

(2) As the cylinder contents cooled, the partial vacuum so formed and the weight of the piston transferred the work to the output shaft on the downward stroke. Exhaust occurred at the end of this stroke.

The piston had to weigh about 70 kg per kW of output, and by its nature the engine size was limited to outputs of a few kilowatts; none the less some 10 000 engines were produced within five years.

At the same time commercial exploitation of oil wells in the USA was occurring, as a result of the pioneer drilling by Drake in 1859. This led to the availability of liquid fuels that were much more convenient to use than gaseous fuels, since these often needed a dedicated gas-producing plant. Liquid fuels without doubt accelerated the development of internal combustion engines, and certainly increased the number of different types available, with oil products providing both the lubricant and the fuel. For the remainder of the 19th century any engine using a gaseous fuel was called a gas engine, and any engine using a liquid was called an oil engine; no reference was necessarily made to the mode of ignition or the different operational principles.

In 1876 the Otto silent engine using the four-stroke cycle was patented and produced. As well as being much quieter than the free-piston engine, the silent engine was about three times as efficient. Otto attributed the improved efficiency to a conjectured stratification of the charge. This erroneous idea was criticised by Sir Dugald Clerk, who appreciated that the improved efficiency was a result of the charge being compressed before ignition. Clerk subsequently provided the first analysis of the Otto cycle (see chapter 2, section 2.2.1).

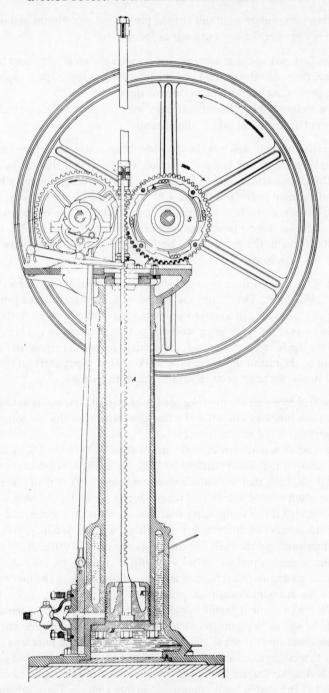

Figure 1.8 Otto and Langen free-piston engine

The concept of compression before ignition can be traced back to Schmidt in 1861, but perhaps more remarkable is the work of Beau de Rochas. As well as advocating the four-stroke cycle, Beau de Rochas included the following points in 1862.

(1) There should be a high volume-to-surface ratio.
(2) The maximum expansion of the gases should be achieved.
(3) The highest possible mixture pressure should occur before ignition.

Beau de Rochas also pointed out that ignition could be achieved by sufficient compression of the charge.

Immediately following the Otto silent engine, two-stroke engines were developed. Patents by Robson in 1877 and 1879 describe the two-stroke cycle with under-piston scavenge, while patents of 1878 and 1881 by Clerk describe the two-stroke cycle with a separate pumping or scavenge cylinder.

The quest for self-propelled vehicles needed engines with better power-to-weight ratios. Daimler was the first person to realise that a light high-speed engine was needed, which would produce greater power by virtue of its higher speed of rotation, 500–1000 rpm. Daimler's patents date from 1884, but his twin cylinder 'V' engine of 1889 was the first to be produced in quantity. By the turn of the century the petrol engine was in a form that would be currently recognisable, but there was still much scope for development and refinement.

The modern compression ignition engine developed from the work of two people, Akroyd Stuart and Rudolf Diesel. Akroyd Stuart's engine, patented in 1890 and first produced in 1892, was a four-stroke compression ignition engine with a compression ratio of about 3 — too low to provide spontaneous ignition of the fuel. Instead, this engine had a large uncooled pre-chamber or vaporiser connected to the main cylinder by a short narrow passage. Initially the vaporiser was heated externally, and the fuel then ignited after it had been sprayed into the vaporiser at the end of the compression stroke. The turbulence generated by the throat to the vaporiser ensured rapid combustion. Once the engine had been started the external heat source could be removed. The fuel was typically a light petroleum distillate such as kerosene or fuel oil; the efficiency of about 15 per cent was comparable with that of the Otto silent engine. The key innovations with the Akroyd Stuart engine were the induction, being solely of air, and the injection of fuel into the combustion chamber.

Diesel's concept of compressing air to such an extent that the fuel would spontaneously ignite after injection was published in 1890, patented in 1892 and achieved in 1893; an early example is shown in figure 1.9. Some of Diesel's aims were unattainable, such as a compression pressure of 240 bar, the use of pulverised coal and an uncooled cylinder. None the less, the prototype ran with an efficiency of 26 per cent, about twice the efficiency of any contemporary power plant and a figure that steam power plant achieved only in the 1930s.

Diesel injected the fuel by means of a high-pressure (70 bar) air blast, since a liquid pump for 'solid' or airless injection was not devised until 1910 by

Figure 1.9   Early (1898) Diesel engine (output: 45 kW at 180 rpm) (reproduced with permission from Singer, *History of Technology*, OUP, 1978)

McKechnie. Air-blast injection necessitated a costly high-pressure air pump and storage vessel; this restricted the use of diesel engines to large stationary and marine applications. Smaller high-speed compression ignition engines were not used for automotive applications until the 1920s. The development depended on experience gained from automotive spark ignition engines, the development of airless quantity-controlled fuel injection pumps by Bosch (chapter 5, section 5.5.2), and the development of suitable combustion systems by people such as Ricardo.

1.3 Characteristics of internal combustion engines

The purpose of this section is to discuss one particular engine — the Ford 'Dover' direct injection in-line six-cylinder truck engine (figure 1.10). This turbocharged engine has a swept volume of 6 litres, weighs 488 kg and produces a power output of 114 kW at 2400 rpm; full details are given in appendix C.

Figure 1.10   Ford Dover compression ignition engine (courtesy of Ford)

In chapter 2 the criteria for judging engine efficiency are derived, along with other performance parameters, such as mechanical efficiency and volumetric efficiency, which provide insight into why a particular engine may or may not be efficient. Thermodynamic cycle analysis also indicates that, regardless of engine type, the cycle efficiency should improve with higher compression ratios. Furthermore, it is shown that for a given compression ratio, cycles that compress a fuel/air mixture have a lower efficiency than cycles that compress pure air. Cycle analysis also indicates the work that can be extracted by an exhaust gas turbine.

The type of fuel required for this engine and the mode of combustion are discussed in chapter 3. Since the fuel is injected into the engine towards the end of the compression stroke, the combustion is not pre-mixed but controlled by diffusion processes − the diffusion of the fuel into the air, the diffusion of the air into the fuel, and the diffusion of the combustion products away from the reaction zone. Turbulence is essential if these processes are to occur in the small time available. The main fuel requirement is that the fuel should readily self-ignite; this is the exact opposite to the requirements for the spark ignition engine. Fuel chemistry and combustion are discussed in chapter 3, along with additives, principally those that either inhibit or promote self-ignition. One of the factors limiting the output of this, and any other, diesel engine is the amount of fuel that can be injected before unburnt fuel leaves the engine as smoke (formed by agglomerated carbon particles). These and other engine emissions are discussed in section 3.8.

A key factor in designing a successful compression ignition engine is the design of the combustion chamber, and the correct matching of the fuel injection to the in-cylinder air motion. These factors are discussed in chapter 5, the counterpart to chapter 4 which discusses spark ignition engines. The design and manufacture of fuel injection equipment is undertaken by specialist manufacturers; however, the final matching of the fuel injection equipment to the engine still has to be done experimentally. This turbocharged engine will have a lower compression ratio than its naturally aspirated counterpart, in order to limit the peak pressures during combustion. One effect of this is that starting is made more difficult; the compression ratio is often chosen to be the minimum that will give reliable starting. None the less, cold weather or poor fuel quality can lead to starting difficulties, and methods to improve starting are discussed in section 5.4.

The induction and exhaust processes are controlled by poppet valves in the cylinder head. The timing of events derives from the camshaft, but is modified by the clearances and the elastic properties of the valve gear. The valve timing for the turbocharged engine will keep the valves open for longer periods than in the naturally aspirated version, since appropriate turbocharger matching can cause the inlet pressure to be greater than the exhaust. These aspects, and the nature of the flow in the inlet and exhaust passages, are discussed in chapter 6. In the turbocharged engine the inlet and exhaust manifold volumes will have been minimised to help reduce the turbocharger response time. In addition, care will have been taken to ensure that the pressure pulses from each cylinder do not interfere with the exhaust system. The design of naturally aspirated induction and exhaust systems can be very involved if the engine performance is to be optimised. Pressure pulses can be reflected as rarefaction waves, and these can be used to improve the induction and exhaust processes. Since these are resonance effects, the engine speed at which maximum benefit occurs depends on the system design, sections 6.3 and 6.4.

The turbocharger is another component that is designed and manufactured by specialists. Matching the turbocharger to the engine is difficult, since the flow characteristics of each machine are fundamentally different. The engine is a slow-running (but large) positive displacement machine, while the turbocharger is a high-speed (but small) non-positive displacement machine, which relies on dynamic flow effects. Turbochargers inevitably introduce a lag when speed or load is increased, since the flow rate can only increase as the rotor speed increases. These aspects, and the design of the compressor and turbine, are covered in chapter 7, along with applications to spark ignition engines. Turbochargers increase the efficiency of compression ignition engines since the power output of the engine is increased more than the mechanical losses.

Some of the mechanical design considerations are dealt with in chapter 8. A six-cylinder in-line engine will have even firing intervals that produce a smooth torque output. In addition, there will be complete balance of all primary and secondary forces and moments that are generated by the reciprocating elements. The increased cycle temperatures in the turbocharged engine make the design and materials selection for the exhaust valve and the piston assembly particularly important. The increased pressures also raise the bearing loads and the role of the lubricant as a coolant will be more important. Computers are increasingly important in design work, as component weights are being reduced and engine outputs are being raised.

The use of computers is increasing in all aspects of engine work: modelling the engine to estimate performance, matching of the fuel injection equipment, selection and performance prediction of the turbocharger, estimation of vehicle performance (speed, fuel consumption etc.) for different vehicles, transmission and usage combinations. Finally, computer-based test systems will have been used for final engine development, and the testing of production engines.

1.4 Additional types of internal combustion engine

Two types of engine that fall outside the simple classification of reciprocating spark ignition or compression ignition engine are the Wankel engine and stratified charge engine.

1.4.1 The Wankel engine

The Wankel engine is a rotary combustion engine, developed from the work of
Felix Wankel. The mode of operation is best explained with reference to figure
1.11. The triangular rotor has a centrally placed internal gear that meshes with a
sun gear that is part of the engine casing. An eccentric that is an integral part of
the output shaft constrains the rotor to follow a planetary motion about the
output shaft. The gear ratios are such that the output shaft rotates at three times
the speed of the rotor, and the tips of the rotor trace out the two-lobe epi-
trochoidal shape of the casing. The compression ratio is dictated geometrically
by the eccentricity of the rotor and the shape of its curved surfaces. The convex
surfaces shown in the diagram maximise and minimise the sealed volumes, to

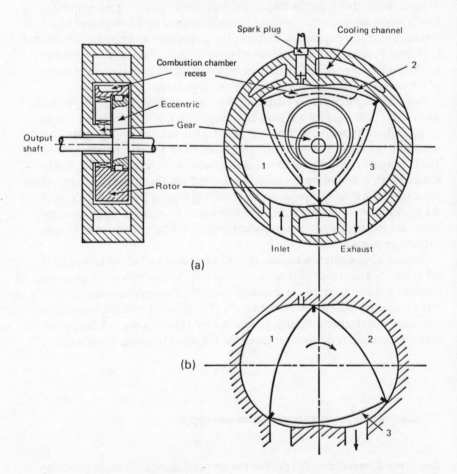

Figure 1.11   The Wankel engine (reproduced with permission from Rogers and
Mayhew (1980a))

give the highest compression ratio and optimum gas exchange. A recess in the combustion chamber provides a better-shaped combustion chamber.

The sequence of events that produces the four-stroke cycle is as follows. In figure 1.11a with the rotor turning in a clockwise direction a charge is drawn into space 1, the preceding charge is at maximum compression in space 2, and the combustion products are being expelled from space 3. When the rotor turns to the position shown in figure 1.11b, space 1 occupied by the charge is at a maximum, and further rotation will cause compression of the charge. The gases in space 2 have been ignited and their expansion provides the power stroke. Space 3 has been reduced in volume, and the exhaust products have been expelled. As in the two-stroke loop or cross scavenge engine there are no valves, and here the gas flow through the inlet and exhaust is controlled by the position of the rotor apex.

For effective operation the Wankel engine requires efficient seals between the sides of the rotor and its casing, and the more demanding requirement of seals at the rotor tips. Additional problems to be solved were cooling of the rotor, the casing around the spark plug and the exhaust passages. Unlike a reciprocating engine, only a small part of the Wankel engine is cooled by the incoming charge. Furthermore, the spark plug had to operate reliably under much hotter conditions. Not until the early 1970s were the sealing problems sufficiently solved for the engine to enter production. The advantages of the Wankel were its compactness, the apparent simplicity, the ease of balance and the potential for high outputs by running at high speeds.

The major disadvantages of the Wankel engine were its low efficiency (caused by limited compression ratios) and the high exhaust emissions resulting from the poor combustion chamber shape. By the mid 1970s concern over firstly engine emissions, and secondly fuel economy led to the demise of the Wankel engine. Experiments with other types of rotary combustion engine have not led to commercial development.

1.4.2 Stratified charge engines

The principle behind stratified charge engines is to have a readily ignitable mixture in the vicinity of the spark plug, and a weaker (normally non-ignitable) mixture in the remainder of the combustion chamber. The purpose of this arrangement is to control the power output of the engine by varying only the fuel supply without throttling the air, thereby eliminating the throttling pressure-drop losses. The stratification of the charge is commonly arranged by division of the combustion chamber to produce a pre-chamber that contains the spark plug. Typically fuel would also be injected into the pre-chamber, so that charge stratification is controlled by the timing and rate of fuel injection. Thus fuel supply is controlled in the same manner as compression ignition engines, yet the ignition timing of the spark controls the start of combustion.

An alternative means of preparing a stratified charge was to provide an extra valve to the pre-chamber, which controlled a separate air/fuel mixture. This was the method used in the Honda CVCC engine, figure 1.12, the first stratified charge engine in regular production.

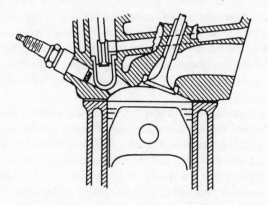

Figure 1.12   Honda CVCC engine, the first in regular production (from Campbell (1978))

The advantages claimed for stratified charge engines include:

(1) Lower exhaust emissions than conventional spark ignition engines, since there is greater control over combustion.
(2) Improved efficiency, since throttling losses are eliminated.
(3) Greater fuel tolerance — suitable fuels should range from petrol to diesel.

Unfortunately stratified charge engines have not met the initial expectations. Compression ignition and spark ignition engines are the result of much development work, and instead of stratified charge engines having the best characteristics of each engine type, they can all too easily end up with the worst characteristics of each engine type, namely:

(1) high exhaust emissions
(2) unimproved efficiency
(3) low power output
(4) high expense (primarily fuel injection equipment and/or extra manufacturing costs).

Since a proper discussion of stratified charge engines requires a knowledge of chapters 2–5, a fuller treatment can be found in appendix D.

1.5 Prospects for internal combustion engines

The future of internal combustion engines will be influenced by two factors: the future cost and availability of suitable fuels, and the development of alternative power plants.

Liquid fuels are by far the most convenient energy source for internal combustion engines, and the majority (over 99 per cent) of such fuels come from crude oil. The oil price is largely governed by political and taxation policy, and there is no reason to suppose that these areas of control will change.

It is very important, but equally difficult, to estimate how long oil supplies will last. Current world consumption of oil is about 65 million barrels per day; this figure is unlikely to rise much before the end of the century owing to improvements in the utilisation of oil and the development of other energy sources. Current known oil reserves would then imply a supply of crude oil for another 25-30 years. However, exploration for oil continues and new reserves are being found; it must also be remembered that oil companies cannot justify expensive exploration work to demonstrate reserves for, say, the next 100 years. An alternative approach is to look at the ratio of oil reserves to the rate of production, see figure 1.13 from Ford (1982). This shows that as the cost of oil has increased, and the cost of prospecting and producing in more difficult fields has risen, then the reserves/production ratio for the Middle East has stabilised at about 50 years. This suggests a continuing equilibrium between supply and demand.

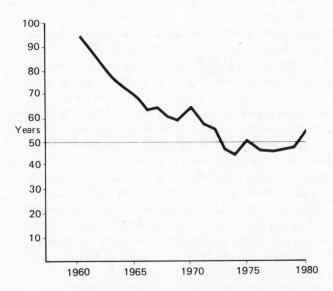

Figure 1.13   Middle East reserves/production ratio (from Ford (1982))

Internal combustion engines can also be fuelled from renewable energy sources. Spark ignition engines run satisfactorily on alcohol-based fuels, and compression ignition engines can operate on vegetable oils. Countries such as Brazil, with no oil reserves but plentiful sources of vegetation, are already operating an alcohol-fuelled policy.

The other major source of hydrocarbons is coal, and even conservative estimates show a 200-year supply from known reserves. One approach is to introduce a suspension of coal particles into the heavy fuel oil used by large compression ignition engines. A more generally applicable alternative is the preparation of fuels by 'liquefaction' or 'gasification' of coal. A comprehensive review of alternative processes is given by Davies (1983), along with the yields of different fuels and their characteristics.

The preceding remarks indicate that the future fuel supply is assured for internal combustion engines, but that other types of power plant may supersede them; the following is a discussion of some of the possibilities.

Steam engines have been used in the past, and would have the advantages of external combustion of any fuel, with readily controlled emissions. However, if it is possible to overcome the low efficiency and other disadvantages it is probable that this would have already been achieved and they would already have been adopted; their future use is thus unlikely.

Stirling engines have been developed, with some units for automotive applications built by United Stirling of Sweden. The fuel economy at full and part load is comparable to compression ignition engines, but the cost of building the complex engine is about 50 per cent greater. As the Stirling engine has external combustion it too has the capability of using a wide range of fuels with readily controlled emissions.

Gas turbines present another alternative: conventionally they use internal combustion, but external combustion is also possible. For efficient operation a gas turbine would require a high efficiency compressor and turbine, high pressure ratio, high combustion temperatures and an effective heat exchanger. These problems have been solved for large aero and industrial gas turbines, but scaling down to even truck engine size changes the design philosophy. The smaller size would dictate the less efficient radial flow compressor, with perhaps a pressure ratio of 5:1 and a regenerative heat exchanger to preserve the efficiency. All these problems could be solved, along with reductions in manufacturing cost, by the development of ceramic materials. However, part load efficiency is likely to remain poor, although this is of less significance in truck applications. The application to private passenger vehicles is even more remote because of the importance of part load efficiency, and the reductions in efficiency that would follow from the smaller size. Marine application has been limited because of the low efficiency. In addition, this can deteriorate rapidly if the turbine blades corrode as a result of the combustion of salt-laden air.

Electric vehicles present an interesting possibility which is currently restricted by the lack of a suitable battery. Lead/acid batteries are widely used, but 1 tonne of batteries only stores the same amount of energy that is available in about 5 litres of fuel. Typical performance figures are still a maximum speed of about 75 km per hour and a range of 75 km. None the less, this would meet the majority of personal transport needs, and is already economically viable for local delivery vehicles.

This leads to the possibility of a hybrid vehicle that has both an internal combustion engine and an electric motor. This is obviously an expensive solution but one that is versatile and efficient by using the motor and/or engine. However, since there is still no real alternative to the lead/acid battery it will be at least a decade before electric vehicles might start to take over from engined vehicles.

2 Thermodynamic Principles

2.1 Introduction and definitions of efficiency

This chapter provides criteria by which to judge the performance of internal combustion engines. Most important are the thermodynamic cycles based on ideal gases undergoing ideal processes. However, internal combustion engines follow a mechanical cycle, not a thermodynamic cycle. The start and end points are mechanically the same in the cycle for an internal combustion engine, whether it is a two-stroke or four-stroke mechanical cycle.

The internal combustion engine is a non-cyclic, open-circuit, quasi steady-flow, work-producing device. None the less it is very convenient to compare internal combustion engines with the ideal air standard cycles, as they are a simple basis for comparison. This can be justified by arguing that the main constituent of the working fluid, nitrogen, remains virtually unchanged in the processes. The internal combustion engine is usually treated as a steady-flow device since most engines are multi-cylindered, with the flow pulsations smoothed at inlet by air filters and at exhaust by silencers.

Air standard cycles have limitations as air and, in particular, air/fuel mixtures do not behave as ideal gases. These effects are discussed at the end of this chapter where computer modelling is introduced. Despite this, the simple air standard cycles are very useful, as they indicate trends. Most important is the trend that as compression ratio increases cycle efficiency should also increase.

At this stage it is necessary to define engine efficiency. This is perhaps the most important parameter to an engineer, although it is often very carelessly defined.

The fuel and air (that is, the reactants) enter the power plant at the temperature T_0 and pressure p_0 of the environment (that is, under ambient conditions). The discharge is usually at p_0, but in general the exhaust products from an internal combustion engine are at a temperature in excess of T_0, the environment temperature. This represents a potential for producing further work if the exhaust products are used as the heat source for an additional cyclic heat power plant. For maximum work production all processes must be reversible and the products must leave the plant at T_0, as well as p_0.

Availability studies show that the work output from an ideal, reversible, non-cyclic, steady-flow, work-producing device is given by

$$W_{REV} = B_{in} - B_{out}$$

where the steady-flow availability function $B \equiv H - T_0 S$.

In the case of an internal combustion engine the ideal case would be when both reactants and products enter and leave the power plant at the temperature and pressure of the environment, albeit with different composition. Thus

$$W_{REV} = (G_{R0} - G_{P0}) \equiv -\Delta G_0 \qquad (2.1)$$

where the Gibbs function G is defined by

$$G \equiv H - TS$$

and

$$G_{R0} = G \text{ of the reactants at } p_0, T_0$$
$$G_{P0} = G \text{ of the products at } p_0, T_0$$

Equation (2.1) defines the maximum amount of work (W_{REV}) that can be obtained from a given chemical reaction. As such it can be used as a basis for comparing the actual output of an internal combustion plant. This leads to a definition of rational efficiency, η_R

$$\eta_R = \frac{W}{W_{REV}} \qquad (2.2)$$

where W is the actual work output.

It is misleading to refer to the rational efficiency as a thermal efficiency, as this term should be reserved for the cycle efficiency of a cyclic device. The upper limit to the rational efficiency can be seen to be 100 per cent, unlike the thermal efficiency of a cyclic device.

This definition is not widely used, since G_0 cannot be determined from simple experiments. Instead calorific value (CV) of a fuel is used

$$CV \equiv (H_{R0} - H_{P0}) \equiv -\Delta H_0 \qquad (2.3)$$

where
$$H_{R0} \equiv H \text{ of the reactants at } p_0, T_0$$
$$H_{P0} \equiv H \text{ of the products at } p_0, T_0$$

and
$$\Delta H_0 \equiv H_{P0} - H_{R0}.$$

The difference in the enthalpies of the products and reactants can be readily found from the heat transfer in a steady-flow combustion calorimeter. This leads to a convenient, but arbitrary, definition of efficiency as

$$\eta_0 \equiv \frac{W}{CV} \equiv \frac{W}{-\Delta H_0} \qquad (2.4)$$

where η_0 = arbitrary overall efficiency
and W = work output.

While the arbitrary overall efficiency is often of the same order as the thermal
efficiency of cyclic plant, it is misleading to refer to it as a thermal efficiency.

Table 2.1, derived from Haywood (1980), shows the difference between
$-\Delta G_0$ and $-\Delta H_0$.

Table 2.1

Fuel	Reaction	$-\Delta G_0$	$-\Delta H_0$ (calorific value)
		MJ/kg of fuel	
C	$C + O_2 \rightarrow CO_2$	32.84	32.77
CO	$CO + \frac{1}{2}O_2 \rightarrow CO_2$	9.19	10.11
H_2	$\begin{cases} H_2 + \frac{1}{2}O_2 \rightarrow H_2O \text{ liq.} \\ H_2 + \frac{1}{2}O_2 \rightarrow H_2O \text{ vap.} \end{cases}$	117.6 113.4	142.0 120.0

The differences in table 2.1 are not particularly significant, especially as the
arbitrary overall efficiency is typically about 30 per cent.

In practice, engineers are more concerned with the fuel consumption of an
engine for a given output rather than with its efficiency. This leads to the use
of specific fuel consumption (sfc), the rate of fuel consumption per unit
power output:

$$\text{sfc} = \frac{\dot{m}_f}{\dot{W}} \text{ kg/J} \tag{2.5}$$

where $\dot{m}_f$ = mass flow rate of fuel
         $\dot{W}$ = power output.

It can be seen that this is inversely proportional to arbitrary overall efficiency,
and is related by the calorific value of the fuel:

$$\text{sfc} = \frac{1}{-\Delta H_0 \, \eta_0} \tag{2.6}$$

It is important also to specify the fuel used. The specific fuel consumption
should be quoted in SI units (kg/MJ), although it is often quoted in metric units
such as (kg/kW h) or in British Units such as (lbs/BHP h).

Sometimes a volumetric basis is used, but this should be avoided as there is a
much greater variation in fuel density than calorific value.

2.2 Ideal air standard cycles

Whether an internal combustion engine operates on a two-stroke or four-stroke cycle and whether it uses spark ignition or compression ignition, it follows a mechanical cycle not a thermodynamic cycle. However, the thermal efficiency of such an engine is assessed by comparison with the thermal efficiency of air standard cycles, because of the similarity between the engine indicator diagram and the state diagram of the corresponding hypothetical cycle. The engine indicator diagram is the record of pressure against cylinder volume, recorded from an actual engine. Pressure/volume diagrams are very useful, as the enclosed area equates to the work in the cycle.

2.2.1 The ideal air standard Otto cycle

The Otto cycle is usually used as a basis of comparison for spark ignition and high-speed compression ignition engines. The cycle consists of four non-flow processes, as shown in figure 2.1. The compression and expansion processes are assumed to be adiabatic (no heat transfer) and reversible, and thus isentropic. The processes are as follows:

1-2   isentropic compression of air through a volume ratio V_2/V_1, the compression ratio r_v
2-3   addition of heat Q_{23} at constant volume
3-4   isentropic expansion of air to the original volume
4-1   rejection of heat Q_{41} at constant volume to complete the cycle.

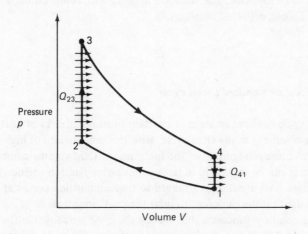

Figure 2.1   Ideal air standard Otto cycle

The efficiency of the Otto cycle, η_{Otto} is

$$\eta_{Otto} = \frac{W}{Q_{23}} = \frac{Q_{23} - Q_{41}}{Q_{23}} = 1 - \frac{Q_{41}}{Q_{23}}$$

By considering air as a perfect gas we have constant specific heat capacities, and for mass m of air the heat transfers are

$$Q_{23} = m\,c_v\,(T_3 - T_2)$$
$$Q_{41} = m\,c_v\,(T_4 - T_1)$$

thus

$$\eta_{Otto} = 1 - \frac{T_4 - T_1}{T_3 - T_2} \tag{2.7}$$

For the two isentropic processes 1–2 and 3–4, $TV^{\gamma-1}$ is a constant. Thus

$$\frac{T_2}{T_1} = \frac{T_3}{T_4} = r_v^{\gamma-1}$$

where γ is the ratio of gas specific heat capacities, c_p/c_v. Thus

$$T_3 = T_4\,r_v^{\gamma-1} \quad \text{and} \quad T_2 = T_1 r_v^{\gamma-1}$$

and substituting into equation (2.7) gives

$$\eta_{Otto} = 1 - \frac{T_4 - T_1}{r_v^{\gamma-1}(T_4 - T_1)} = 1 - \frac{1}{r_v^{\gamma-1}} \tag{2.8}$$

The value of η_{Otto} depends on the compression ratio, r_v, and not the temperatures in the cycle. To make a comparison with a real engine, only the compression ratio needs to be specified. The variation in η_{Otto} with compression ratio is shown in figure 2.2 along with that of η_{Diesel}.

2.2.2 The ideal air standard Diesel cycle

The Diesel cycle has heat addition at constant pressure, instead of heat addition at constant volume as in the Otto cycle. With the combination of high compression ratio, to cause self-ignition of the fuel, and constant-volume combustion the peak pressures can be very high. In large compression ignition engines, such as marine engines, fuel injection is arranged so that combustion occurs at approximately constant pressure in order to limit the peak pressures.

The four non-flow processes constituting the cycle are shown in the state diagram (figure 2.3). Again, the best way to calculate the cycle efficiency is to calculate the temperatures around the cycle. To do this it is necessary to specify

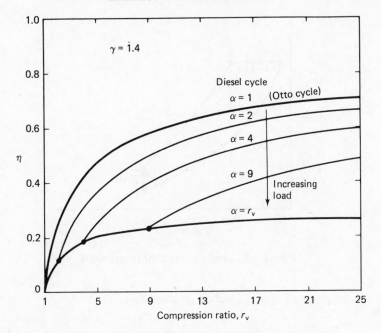

Figure 2.2   Diesel cycle efficiency for different cut-off ratios, α

the cut-off ratio or load ratio:

$$\alpha \equiv V_3/V_2$$

The processes are all reversible, and are as follows:

1-2   isentropic compression of air through a volume ratio V_2/V_1, the compression ratio r_v

2-3   addition of heat Q_{23} at constant pressure while the volume expands through a ratio V_3/V_2, the cut-off ratio α

3-4   isentropic expansion of air to the original volume

4-1   rejection of heat Q_{41} at constant volume to complete the cycle.

The efficiency of the Diesel cycle, η_{Diesel}, is

$$\eta_{Diesel} = \frac{W}{Q_{23}} = \frac{Q_{23} - Q_{41}}{Q_{23}} = 1 - \frac{Q_{41}}{Q_{23}}$$

By treating air as a perfect gas we have constant specific heat capacities, and for mass m of air the heat transfers are

$$Q_{23} = m \, c_p \, (T_3 - T_2)$$
$$Q_{41} = m \, c_v \, (T_4 - T_1)$$

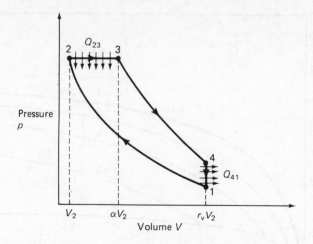

Figure 2.3   Ideal air standard Diesel cycle

Note that the process 2-3 is at constant pressure, thus

$$\eta_{\text{Diesel}} = 1 - \frac{1}{\gamma} \frac{T_4 - T_1}{T_3 - T_2} \tag{2.9}$$

For the isentropic process 1-2, $TV^{\gamma-1}$ is a constant:

$$T_2 = T_1 \, r_v^{\gamma-1}$$

For the constant pressure process 2-3

$$\frac{T_3}{T_2} = \frac{V_3}{V_2} = \alpha \quad \text{thus} \quad T_3 = \alpha r_v^{\gamma-1} T_1$$

For the isentropic process 3-4, $TV^{\gamma-1}$ is a constant:

$$\frac{T_4}{T_3} = \left(\frac{V_3}{V_4}\right)^{\gamma-1} = \left(\frac{\alpha}{r_v}\right)^{\gamma-1}$$

thus

$$T_4 = \left(\frac{\alpha}{r_v}\right)^{\gamma-1} T_3 = \alpha r_v^{\gamma-1} \left(\frac{\alpha}{r_v}\right)^{\gamma-1} T_1 = \alpha^\gamma T_1$$

Substituting for all the temperatures in equation (2.9) in terms of T_1 gives

$$\eta_{\text{Diesel}} = 1 - \frac{1}{\gamma} \cdot \frac{\alpha^\gamma - 1}{\alpha r_v^{\gamma-1} - r_v^{\gamma-1}} = 1 - \frac{1}{r_v^{\gamma-1}} \left[\frac{\alpha^\gamma - 1}{\gamma(\alpha - 1)}\right] \tag{2.10}$$

At this stage it is worth making a comparison between the air standard Otto cycle efficiency (equation 2.8) and the air standard Diesel cycle efficiency (equation 2.10).

The Diesel cycle efficiency is less convenient; it is not solely dependent on compression ratio, r_v, but is also dependent on the cut-off ratio α. The two expressions are the same, except for the term in square brackets

$$\left[\frac{\alpha^\gamma - 1}{\gamma(\alpha - 1)} \right]$$

The cut-off ratio lies in the range $1 < \alpha < r_v$, and is thus always greater than unity. Consequently the expression in square brackets is always greater than unity, and the Diesel cycle efficiency is less than the Otto cycle efficiency *for the same compression ratio*. This is shown in figure 2.2 where efficiencies have been calculated for a variety of compression ratios and cut-off ratios. There are two limiting cases. The first is, as $\alpha \to 1$, then $\eta_{\text{Diesel}} \to \eta_{\text{Otto}}$. The second limiting case is when $\alpha \to r_v$ and point $3 \to 4$ in the cycle, and the expansion is wholly at constant pressure; this corresponds to maximum work output in the cycle. Figure 2.2 also shows that as load increases, with a fixed compression ratio the efficiency reduces. The compression ratio of a compression ignition engine is usually greater than for a spark ignition engine, so the former is usually more efficient.

2.2.3 The ideal air standard Dual cycle

In practice, combustion occurs neither at constant volume nor at constant pressure. This leads to the Dual, Limited Pressure, or Mixed cycle which has heat addition in two stages, firstly at constant volume, and secondly at constant pressure. The state diagram is shown in figure 2.4; again all processes are assumed to be reversible. As might be expected, the efficiency lies between that of the Otto cycle and the Diesel cycle. An analysis can be found in several sources such as Taylor (1966) or Benson and Whitehouse (1979), but is not given here as the extra complication does not give results significantly closer to reality.

2.2.4 The ideal air standard Atkinson cycle

This is commonly used to describe any cycle in which the expansion stroke is greater than the compression stroke. Figure 2.5 shows the limiting case for the Atkinson cycle in which expansion is down to pressure p_1. All processes are reversible, and processes 1–2 and 3–4 are also adiabatic.

The shaded area (1A4) represents the increased work (or reduced heat rejection) when the Atkinson cycle is compared to the Otto cycle. The mechanical difficulties of arranging unequal compression and expansion strokes have prevented the development of engines working on the Atkinson cycle. However, expansion A4 can be arranged in a separate cycle, for example an exhaust turbine. This subject is treated more fully in chapter 7.

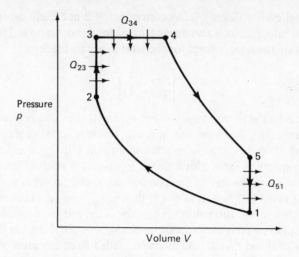

Figure 2.4   Ideal air standard Dual cycle

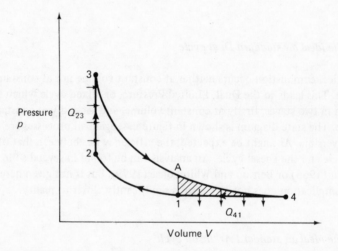

Figure 2.5   Ideal air standard Atkinson cycle

2.3 Comparison between thermodynamic and mechanical cycles

Internal combustion engines operate on a mechanical cycle, not a thermodynamic cycle. Although it is an arbitrary procedure it is very convenient to
compare the performance of real non-cyclic engines with thermal efficiencies of

hypothetical cycles. This approach arises from the similarity of the engine indicator diagram, figure 2.6, and the state diagram of a hypothetical cycle, figure 2.1 or figure 2.3.

Figure 2.6 is a stylised indicator diagram for a high-speed four-stroke engine, with exaggerated pressure difference between induction and exhaust strokes. As before, r_v is the compression ratio, and V_c is the clearance volume with the piston at top dead centre (tdc). The diagram could be from either a spark ignition or a compression ignition engine, as in both cases combustion occurs neither at constant pressure nor at constant volume. For simplicity it can be idealised as constant-volume combustion, figure 2.7, and then compared with the Otto cycle, figure 2.1.

In the idealised indicator diagram, induction 0–1 is assumed to occur with no pressure drop. The compression and expansion (1–2, 3–4) are not adiabatic, so neither are they isentropic. Combustion is assumed to occur instantaneously at constant volume, 2–3. Finally, when the exhaust valve opens blow-down occurs instantaneously, with the exhaust expanding into the manifold 4–1, and the exhaust stroke occurring with no pressure drop, 1–0. The idealised indicator diagram is used as a basis for the simplest computer models.

In comparison, the Otto cycle assumes that

(1) air behaves as a perfect gas with constant specific heat capacity, and all processes are fully reversible
(2) there is no induction or exhaust process, but a fixed quantity of air and no leakage
(3) heat addition is from an external source, in contrast to internal combustion
(4) heat rejection is to the environment to complete the cycle, as opposed to blow-down and the exhaust stroke.

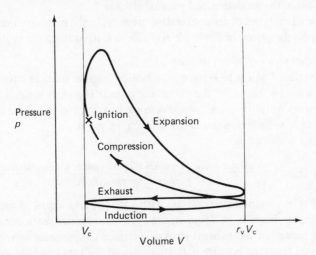

Figure 2.6 Stylised indicator diagram for four-stroke engine

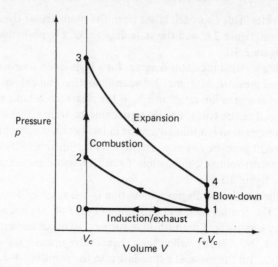

Figure 2.7 Idealised indicator diagram for four-stroke engine

2.4 Additional performance parameters for internal combustion engines

The rational efficiency and arbitrary overall efficiency have already been defined in section 2.1. The specific fuel consumption which has a much greater practical significance has also been defined. The additional parameters relate to the work output per unit swept volume in terms of a mean effective pressure, and the effectiveness of the induction and exhaust strokes.

There are two types of mean effective pressure, based on either the work done by the gas on the piston or the work available as output from the engine.

(a) Indicated mean effective pressure (imep)

The area enclosed on the p-V trace or indicator diagram from an engine is the indicated work (W_I) done by the gas on the piston. The imep is a measure of the indicated work output per unit swept volume, in a form independent of the size and number of cylinders in the engine and engine speed.

The imep is defined as

$$\text{imep (N/m}^2\text{)} = \frac{\text{indicated work output (N m) per cylinder per mechanical cycle}}{\text{swept volume per cylinder (m}^3\text{)}} \quad (2.11)$$

Figure 2.8 shows an indicator diagram with a shaded area, equal to the net area of the indicator diagram. In a four-stroke cycle the negative work occurring during the induction and exhaust strokes is termed the *pumping loss*, and has to be subtracted from the positive indicated work of the other two strokes. When an engine is throttled the pumping loss increases, thereby reducing the engine

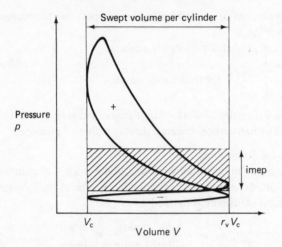

Figure 2.8 Indicated mean effective pressure

efficiency. In figure 2.8 the shaded area has the same volume scale as the indicator diagram, so the height of the shaded area must correspond to the imep.

The imep bears no relation to the peak pressure in an engine, but is a characteristic of engine type. The imep in naturally aspirated four-stroke spark ignition engines will be smaller than the imep of a similar turbocharged engine. This is mainly because the turbocharged engine has greater air density at the start of compression, so more fuel can be burnt.

(b) *Brake mean effective pressure (bmep)*

The work output of an engine, as measured by a brake or dynamometer, is more important than the indicated work output. This leads to a definition of bmep, $\overline{p}_b$, very similar to equation (2.11):

$$\overline{p}_b(N/m^2) = \frac{\text{brake work output (N m) per cylinder per mechanical cycle}}{\text{swept volume per cylinder (m}^3)}$$

(2.12)

or in terms of the engine brake power

$$\text{brake power} = \overline{p}_b \, L \, A \, N'$$
$$= \overline{p}_b \, (L \, A \, n)N^* = p_b \, V_s \, N^* \tag{2.13}$$

where L = piston stroke (m)
        A = piston area (m^2)
        n = number of cylinders
        V_s = engine swept volume (m^3)

and N' = number of mechanical cycles of operation per second
        $N^* = N'/n$

$$= \begin{cases} \text{rev./s for two-stroke engines} \\ \dfrac{\text{rev./s}}{2} \text{ for four-stroke engines.} \end{cases}$$

The bmep is a measure of work output from an engine, and not of pressures in the engine. The name arises because its unit is that of pressure.

(c) *Mechanical efficiency, η_m*
The difference between indicated work and brake work is accounted for by friction, and work done in driving essential items such as the lubricating oil pump. Mechanical efficiency is defined as

$$\eta_m = \frac{\text{brake power}}{\text{indicated power}} = \frac{\text{bmep}}{\text{imep}} \tag{2.14}$$

(d) *Indicated efficiency, η_i*
When comparing the performance of engines it is sometimes useful to isolate the mechanical losses. This leads to the use of indicated (arbitrary overall) efficiency as a means of examining the thermodynamic processes in an engine:

$$\eta_i = \frac{\dot{W}_i}{\dot{m}_f.CV} = \frac{\dot{W}}{\dot{m}_f.CV.\eta_m} = \frac{\eta_o}{\eta_m} \tag{2.15}$$

(e) *Volumetric efficiency, η_v*
Volumetric efficiency is a measure of the effectiveness of the induction and exhaust processes. Even though some engines inhale a mixture of fuel and air it is convenient, but arbitrary, to define volumetric efficiency as

$$\eta_v = \frac{\text{mass of air inhaled per cylinder per cycle}}{\text{mass of air to occupy swept volume per cylinder at ambient } p \text{ and } T}$$

Assuming air obeys the Gas Laws, this can be rewritten as

$$\eta_v = \frac{\text{volume of ambient density air inhaled per cylinder per cycle}}{\text{cylinder volume}}$$

$$= \frac{\dot{V}_a}{V_s N^*} \tag{2.16}$$

where $\dot{V}_a$ = volumetric flow rate of air with ambient density
        V_s = engine swept volume
        $$N^* = \begin{cases} \text{rev./s for two-stroke engines} \\ \dfrac{\text{rev./s}}{2} \text{ for four-stroke engines} \end{cases}$$

In the case of supercharged engines, compressor delivery conditions should be used instead of ambient conditions. Volumetric efficiency has a direct effect on power output, as the mass of air in a cylinder determines the amount of fuel that can be burnt. In a well-designed, naturally aspirated engine the volumetric efficiency can be over 90 per cent.

Volumetric efficiency depends on the density of the gases at the end of the induction process; this depends on the temperature and pressure of the charge. There will be pressure drops in the inlet passages and at the inlet valve owing to viscous effects. The charge temperature will be raised by heat transfer from the induction manifold, mixing with residual gases, and heat transfer from the piston, valves and cylinder. In a petrol engine, fuel evaporation can cool the charge by as much as 25 K, and alcohol fuels have much greater cooling effects; this improves volumetric efficiency.

In an idealised process, with charge and residuals having the same specific heat capacity and molar mass, the temperature of the residual gases does not affect volumetric efficiency. This is because in the idealised process induction and exhaust occur at the same constant pressure, and when the two gases mix the contraction on cooling of the residual gases is exactly balanced by the expansion of the charge.

In practice, induction and exhaust processes do not occur at the same pressure and the effect this has on volumetric efficiency for different compression ratios is discussed by Taylor (1966).

2.5 Fuel–air cycle

The simple ideal air standard cycles overestimate the engine efficiency by a factor of about 2. A significant simplification in the air standard cycles is the assumption of constant specific heat capacities. Heat capacities of gases are strongly temperature-dependent, as shown by figure 2.9.

The molar constant-volume heat capacity will also vary, as will γ the ratio of heat capacities:

$$C_p - C_v = R_0, \quad \gamma = C_p/C_v$$

If this is allowed for, air standard Otto cycle efficiency falls from 57 per cent to 49.4 per cent for a compression ratio of 8.

When allowance is made for the presence of fuel and combustion products, there is an even greater reduction in cycle efficiency. This leads to the concept of a fuel–air cycle which is the same as the ideal air standard Otto cycle, except that allowance is made for the real thermodynamic behaviour of the gases. The cycle assumes instantaneous complete combustion, no heat transfer, and reversible compression and expansion. Taylor (1966) discusses these matters in detail and

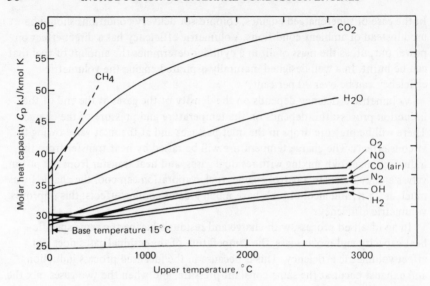

Figure 2.9   Molar heat capacity at constant pressure of gases above 15°C quoted as averages between 15°C and abscissa temperature (adapted from Taylor (1966))

provides results in graphical form. Figures 2.10 and 2.11 show the variation in fuel–air cycle efficiency as a function of equivalence ratio for a range of compression ratios. Equivalence ratio ϕ is defined as the chemically correct (stoichiometric) air/fuel ratio divided by the actual air/fuel ratio. The datum conditions at the start of the compression stroke are pressure (p_1) 1.013 bar, temperature (T_1) 115°C, mass fraction of combustion residuals (f) 0.05, and specific humidity (ω) 0.02 – the mass fraction of water vapour.

The fuel 1-octene has the formula $C_8 H_{16}$, and structure

$$\begin{array}{c} H \\ \\ H \end{array}\!\!\!\!C = \overset{H}{\underset{H}{C}} - \overset{H}{\underset{H}{C}} - \overset{H}{\underset{H}{C}} - \overset{H}{\underset{H}{C}} - \overset{H}{\underset{H}{C}} - \overset{H}{\underset{H}{C}} - \overset{H}{\underset{H}{C}} - H$$

Figure 2.10 shows the pronounced reduction in efficiency of the fuel–air cycle for rich mixtures. The improvement in cycle efficiency with increasing compression ratio is shown in figure 2.11, where the ideal air standard Otto cycle efficiency has been included for comparison.

In order to make allowances for the losses due to phenomena such as heat transfer and finite combustion time, it is necessary to develop computer models.

Prior to the development of computer models, estimates were made for the various losses that occur in real operating cycles. Again considering the Otto

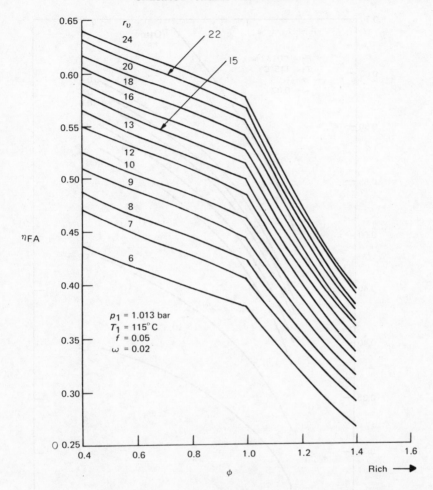

Figure 2.10　Variation of efficiency with equivalence ratio for a constant-volume fuel–air cycle with 1-octene fuel for different compression ratios (adapted from Taylor (1966))

cycle, these are as follows:

(a) 'Finite piston speed losses' occur since combustion takes a finite time and cannot occur at constant volume. This leads to the rounding of the indicator diagram and Taylor (1966) estimates these losses as being about 6 per cent.

(b) 'Heat losses', in particular between the end of the compression stroke and the beginning of the expansion stroke. Estimates of up to 12 per cent have been made by both Taylor (1966) and Ricardo and Hempson (1968). How-

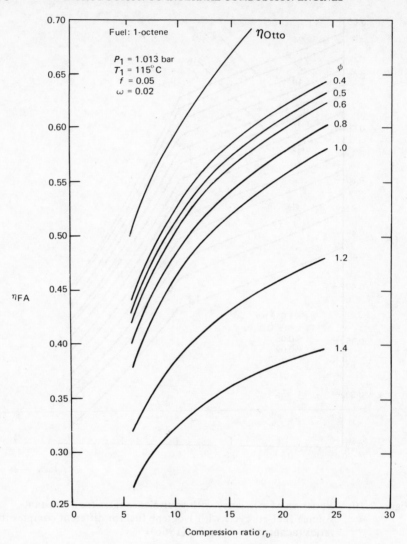

Figure 2.11   Variation of efficiency with compression ratio for a constant-volume fuel–air cycle with 1-octene fuel for different equivalence ratios (adapted from Taylor (1966))

ever, with no heat transfer the cycle temperatures would be raised and the fuel–air cycle efficiencies would be reduced slightly because of increasing gas specific heats and dissociation.

(c) Exhaust losses due to the exhaust valve opening before the end of the expansion stroke. This promotes gas exchange but reduces the expansion work. Taylor (1966) estimates these losses as 2 per cent.

Since the fuel is injected towards the end of the compression stroke in compression ignition engines (unlike the spark ignition engine where it is pre-mixed with the air) the compression process will be closer to ideal in the compression ignition engine than in the spark ignition engine. This is another reason for the better fuel economy of the compression ignition engine.

2.6 Computer models

In internal combustion engines the induction, compression, expansion and exhaust strokes are all influenced by the combustion process. In any engine model it is necessary to include all processes, of which combustion is the most complex. Combustion models are discussed in chapter 3, section 3.9.

Benson and Whitehouse (1979) provide a useful introduction to engine modelling by giving Fortran programs for spark ignition and compression ignition engine cycle calculations. The working of the programs is explained and typical results are presented. These models have now been superseded but none the less they provide a good introduction.

The use of engine models is increasing as engine testing becomes more expensive and computing becomes cheaper. Additionally, once an engine model has been set up, results can be produced more quickly. However, it is still necessary to check model results against engine results to calibrate the model. Engine manufacturers and research organisations either develop their own models or buy-in the expertise from specialists.

The aspects of engine models discussed in this section are the compression and expansion processes. The main difference between spark ignition and compression ignition cycles is in the combustion process. The other significant difference is that spark ignition engines usually induct and compress a fuel/air mixture.

This section considers the compression and expansion strokes in the approach adopted by Benson and Whitehouse (1979).

The 1st Law of Thermodynamics expressed in differential form is

$$dQ = dE + dW \qquad (2.17)$$

The heat transfer, dQ, will be taken as zero in this simple model. Heat transfer in internal combustion engines is still very poorly understood, and there is a shortage of experimental data. A widely used correlation for heat transfer inside an engine cylinder is due to Annand (1963); this allows for convection and radiation, and is discussed by Benson and Whitehouse (1979).

The change in absolute energy of the cylinder contents, dE, is a complex function of temperature, which arises because of the variation of the gas specific heat capacities with temperature. Equation (2.17) cannot be solved analytically, so a numerical solution is used, which breaks each process into a series of steps. Consider the ith to the $(i + 1)^{th}$ steps, for which the values of E can be found as functions of the temperatures

$$dE = E(T_{i+1}) - E(T_i)$$

The work term, dW, equals pdV for an infinitesimal change. If the change is sufficiently small the work term can be approximated by

$$dW = \frac{(p_i + p_{i+1})}{2} (V_{i+1} - V_i)$$

These results can be substituted into equation (2.17) to give

$$0 = E(T_{i+1}) - E(T_i) + \frac{(p_{i+1} + p_i)}{2} (V_{i+1} - V_i) \qquad (2.18)$$

To find p_{i+1}, the state law is applied

$$pV = nR_0T$$

If the gas composition is unchanged, n the number of moles is constant, thus

$$p_{i+1} = \frac{(V_i)}{(V_{i+1})} \frac{(T_{i+1})}{(T_i)} p_i$$

For each step change in volume the temperature (T_{i+1}) can be found, but because of the complex nature of equation (2.18) an iterative solution is needed. The Newton–Raphson method is used because of its rapid convergence.

For the $(n + 1)^{th}$ iteration

$$(T_{i+1})_{n+1} = (T_{i+1})_n - \frac{f(E)_n}{f'(E)_n} \qquad (2.19)$$

where

$$f'(E) = \frac{d}{dT} (f(E)).$$

For each volume step the first estimate of T_{i+1} is made by assuming an isentropic process with constant specific heat capacities calculated at temperature T_i.

This model is equivalent to that used in section 2.5 on the fuel-air cycle; the efficiencies quoted from this model are slightly higher than those of Taylor (1966). The differences can be attributed to Taylor considering starting conditions of 115°C with 5 per cent exhaust residuals and 2 per cent water vapour, and fuel of octene (C_8H_{16}) as opposed to octane (C_8H_{18}). As an example, figure 2.12 shows the variation of efficiency with equivalence ratio for a compression ratio of 8 using octane as fuel; this can be compared with figure 2.10.

The simple computer model is comparable to the results from the fuel-air cycle as no account is taken of either heat transfer or finite combustion times.

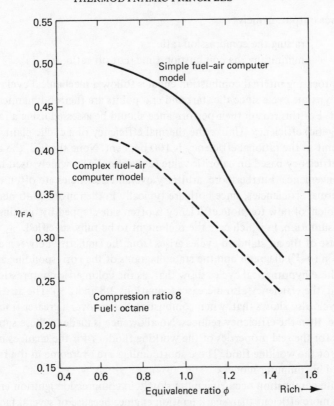

Figure 2.12   Variation of efficiency with equivalence ratio for simple fuel–air
computer model and complex fuel–air model with allowance for
heat transfer and combustion time (reprinted with permission
from Benson and Whitehouse (1979), © Pergamon Press Ltd)

More complex computer models have not been discussed here, but make allowance for finite combustion time, heat transfer and reaction rate kinetics.
Complex models have close agreement to results that might be obtained from an
experimental determination of indicated arbitrary overall efficiency. This is not
surprising, as the complex models will have empirical constants derived from
experiments to calibrate the model.

2.7 Conclusions

This chapter has devised criteria by which to judge the performance of actual
engines, and it has also identified some of the means of increasing both efficiency

and power output, namely:

> raising the compression ratio
> minimising the combustion time (cut-off ratio, $\alpha \to 1$)

Reciprocating internal combustion engines follow a mechanical cycle — not a thermodynamic cycle since the start and end points are thermodynamically different. For this reason their performance should be assessed using a rational (or exergetic) efficiency. Unlike the thermal efficiency of a cyclic plant, the upper limit to the rational efficiency is 100 per cent. None the less, the arbitrary overall efficiency based on calorific value (equation 2.4) is widely used because of its convenience. Furthermore, arbitrary overall efficiencies are often compared with thermal efficiencies, since both are typically in the range 30–40 per cent. The problem of how to quote efficiency is often side-stepped by quoting specific fuel consumption, in which case the fuel ought to be fully specified.

The use of the air standard cycles arises from the similarity between engine indication (p-V) diagrams and the state diagrams of the corresponding air standard cycle. These hypothetical cycles show that, as the volumetric compression ratio is increased, the efficiency also increases (equations 2.8 and 2.9). The air standard Diesel cycle also shows that, when combustion occurs over a greater fraction of the cycle, then the efficiency reduces. No allowance is made in these simple analyses for the real properties of the working fluid, or for the changes in composition of the working fluid. These shortcomings are overcome in the fuel–air cycle and in computer models.

Despite combustion occurring more slowly in compression ignition engines, they are more efficient than spark ignition engines because of several factors:

(1) their higher compression ratios
(2) their power output is controlled by the quantity of fuel injected, not by throttling, with its associated losses
(3) during compression the behaviour of air is closer to ideal than the behaviour of a fuel/air mixture.

Additional performance parameters have also been defined: imep, bmep, mechanical efficiency, indicated efficiency and volumetric efficiency. The indicated performance parameters are particularly useful since they measure the thermodynamic performance, as distinct from brake performance which includes the associated mechanical losses.

2.8 Examples

Example 2.1

(i) The Rolls Royce CV12 turbocharged four-stroke direct injection diesel engine
has a displacement of 26.1 litres. The engine has a maximum output of
900 kW at 2300 rpm and is derated to 397.5 kW at 1800 rpm for industrial
use. What is the bmep for each of these types?
(ii) The high-performance version of the CV12 has an sfc of 0.063 kg/MJ at
maximum power, and a minimum sfc of 0.057 kg/MJ. Calculate the arbitrary
overall efficiencies for both conditions and the fuel flow rate at maximum
power. The calorific value of the fuel is 42 MJ/kg.

Solution:

(i) Using equation (2.13)

$$\text{brake power} = \bar{p}_b V_s N^*$$

remembering that for a four-stroke engine $N^* = (\text{rev./s})/2$.
Rearranging gives

$$\text{bmep, } \bar{p}_b = \frac{\text{brake power}}{V_s \dfrac{\text{rev./s}}{2}}$$

For the high-performance engine

$$\bar{p}_b = \frac{900 \times 10^3}{26.1 \times 10^{-3} \times (2300/120)} = 18.0 \times 10^5 \text{ N/m}^2$$

$$= \underline{18 \text{ bar}}$$

For the industrial engine

$$\bar{p}_b = \frac{397.5 \times 10^3}{26.1 \times 10^{-3} \times (1800/120)} = 10.15 \times 10^5 \text{ N/m}^2$$

$$= \underline{10.15 \text{ bar}}$$

(ii) Equation (2.6) relates specific fuel consumption to arbitrary overall
efficiency:

$$\text{sfc} = \frac{1}{-\Delta H_0 \eta_0} \quad \text{or} \quad \eta_0 = \frac{1}{\text{CV.sfc}}$$

At maximum power

$$\eta_0 = \frac{1}{0.063 \times 42} = 37.8 \text{ per cent}$$

At maximum economy

$$\eta_0 = \frac{1}{0.057 \times 42} = 41.8 \text{ per cent}$$

Finally calculate the maximum flow rate of fuel, $\dot{m}_f$:

$$\dot{m}_f = \text{brake power. sfc}$$

$$= (900 \times 10^3) \times (0.063 \times 10^{-6}) = 0.0567 \text{ kg/s}$$

Example 2.2

A high-performance four-stroke SI engine has a swept volume of 875 cm³ and a compression ratio of 10:1. The indicated efficiency is 55 per cent of the corresponding ideal air standard Otto cycle. At 8000 rpm, the mechanical efficiency is 85 per cent, and the volumetric efficiency is 90 per cent. The air/fuel ratio (gravimetric, that is, by mass) is 13:1 and the calorific value of the fuel is 44 MJ/kg. The air is inducted at 20°C and 1 bar.

Calculate: (i) the arbitrary overall efficiency and the sfc

(ii) the air mass flow rate, power output and bmep

(i) The first step is to find the arbitrary overall efficiency, η_0: $\eta_0 = \eta_m.\eta_i$, and the question states that $\eta_i = 0.55 \, \eta_{Otto}$

$$\eta_{Otto} = 1 - \frac{1}{r_v^{\gamma-1}} \tag{2.8}$$

$$= 1 - \frac{1}{10^{(1.4-1)}} = 0.602$$

Thus $\eta_0 = 0.85 \times 0.55 \times 0.602 = \underline{28.1 \text{ per cent}}$

From equation (2.6)

$$\text{sfc} = \frac{1}{CV.\eta_0} = \frac{1}{44 \times 0.281} = \underline{0.0809 \text{ kg/MJ}}$$

(ii) The air mass flow rate is found from the volume flow rate of air using the equation of state.

$$p\dot{V}_a = m_a R_a T, \qquad R_a = 287 \text{ J/kg K}$$

From equation (2.16) $\dot{V}_a = V_s \, \eta_v N^*$, where $N^* = (\text{rev./s})/2$ for a four-stroke engine. Combining and rearranging gives

$$\dot{m}_a = \frac{p V_s \eta_v N^*}{R_a T} = \frac{10^5 \times 875 \times 10^{-6} \times 0.9 \times (8000/120)}{287 \times 293}$$

$$= 0.0624 \text{ kg/s}$$

The power output can be found for the fuel flow rate, since the efficiency (or sfc) is known. The fuel flow rate is then found for the air flow rate, from the air/fuel ratio (A/F).

$$\dot{m}_f = \dot{m}_a / A/F$$

Brake power output = $\dfrac{\dot{m}_f}{sfc}$ = $\dfrac{0.0624}{13.0809 \times 10^{-6}}$ = $\underline{59.3 \text{ kW}}$

Finally, from equation (2.13)

$$\bar{p}_b = \frac{\dot{W}_b}{V_s N^*} = \frac{59.3 \times 10^3}{875 \times 10^{-6} \times (8000/120)} = \underline{10.2 \text{ bar}}$$

2.9 Problems

2.1 For the ideal air standard Diesel cycle with a volumetric compression ratio of 17:1 calculate the efficiencies for cut-off rates of 1, 2, 4, 9. Take $\gamma = 1.4$. The answers can be checked with figure 2.2.

2.2 Outline the shortcomings of the simple ideal cycles, and explain how the fuel–air cycle and computer models overcome these problems.

2.3 A 2 litre four-stroke indirect injection diesel engine is designed to run at 4500 rpm with a power output of 45 kW; the volumetric efficiency is found to be 80 per cent. The sfc is 0.071 kg/MJ and the fuel has a calorific value of 42 MJ/kg. The ambient conditions for the test were 20°C and 1 bar. Calculate the bmep, the arbitrary overall efficiency, and the air/fuel ratio.

2.4 A twin-cylinder two-stroke engine has a swept volume of 150 cm³. The maximum power output is 19 kW at 11 000 rpm. At this condition the sfc is 0.11 kg/MJ, and the gravimetric air/fuel ratio is 12:1. If ambient test conditions were 10°C and 1.03 bar, and the fuel has a calorific value of 44 MJ/kg, calculate: the bmep, the arbitrary overall efficiency and the volumetric efficiency.

2.5 A four-stroke 3 litre V6 spark ignition petrol engine has a maximum power output of 100 kW at 5500 rpm, and a maximum torque of 236 N m at 3000 rpm. The minimum sfc is 0.090 kg/MJ at 3000 rpm, and the air flow rate is 0.068 m³/s. The compression ratio is 8.9:1 and the mechanical efficiency is 90 per cent. The engine was tested under ambient conditions of 20°C and 1 bar; take the calorific value of the fuel to be 44 MJ/kg.
(a) Calculate the power output at 3000 rpm and the torque output at 5500 rpm.

(b) Calculate for both speeds the bmep and the imep.
(c) How does the arbitrary overall efficiency at 3000 rpm compare with the corresponding air standard Otto cycle efficiency?
(d) What is the volumetric efficiency and air/fuel ratio at 3000 rpm?

3 Combustion and Fuels

3.1 Introduction

The fundamental difference between spark ignition (SI) and compression ignition (CI) engines lies in the type of combustion that occurs, and not in whether the process is idealised as an Otto cycle or a Diesel cycle. The combustion process occurs at neither constant volume (Otto cycle), nor constant pressure (Diesel cycle). The difference between the two combustion processes is that spark ignition engines usually have pre-mixed flames while compression ignition engines have diffusion flames. With pre-mixed combustion the fuel/air mixture must always be close to stoichiometric (chemically correct) for reliable ignition and combustion. To control the power output a spark ignition engine is throttled, thus reducing the mass of fuel and air in the combustion chamber; this reduces the cycle efficiency. In contrast, for compression ignition engines with fuel injection the mixture is close to stoichiometric only at the flame front. The output of compression ignition engines can thus be varied by controlling the amount of fuel injected; this accounts for their superior part load fuel economy.

With pre-mixed reactants the flame moves relative to the reactants, so separating the reactants and products. An example of pre-mixed combustion is with oxy-acetylene equipment; for welding, the flame is fuel-rich to prevent oxidation of the metal, while, for metal cutting, the flame is oxygen-rich in order to burn as well as to melt the metal.

With diffusion flames, the flame occurs at the interface between fuel and oxidant. The products of combustion diffuse into the oxidant, and the oxidant diffuses through the products. Similar processes occur on the fuel side of the flame, and the burning rate is controlled by diffusion. A common example of a diffusion flame is the candle. The fuel is melted and evaporated by radiation from the flame, and then oxidised by air; the process is evidently one governed by diffusion as the reactants are not pre-mixed.

The Bunsen burner, shown in figure 3.1, has both a pre-mixed flame and a diffusion flame. The air entrained at the base of the burner is not sufficient for complete combustion with a single pre-mixed flame. Consequently, a second

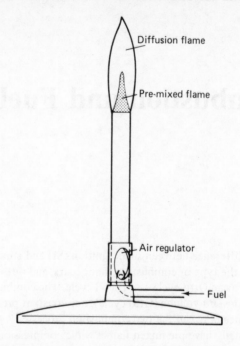

Figure 3.1   Bunsen burner with pre-mixed and diffusion flames

flame front is established at the interface where the air is diffusing into the unburnt fuel.

The physics and chemistry of combustion are covered in some detail by both
Gaydon and Wolfhard (1979) and Lewis and von Elbe (1961), but neither book
devotes much attention to combustion in internal combustion engines. Hydro-
carbon/air mixtures have a maximum laminar burning velocity of about 0.5 m/s,
a notable exception being acetylene/air with a value of 1.58 m/s.

An order of magnitude calculation for the burning time in a cylinder of
100 mm diameter with central ignition gives about 100 ms. However, for an
engine running at 3000 rpm the combustion time can only be about 10 ms.
This shows the importance of turbulence in speeding combustion by at least
an order of magnitude.

Turbulence is generated as a result of the induction and compression processes,
and the geometry of the combustion chamber. In addition there can be an ordered
air motion such as swirl which is particularly important in diesel engines. This is
obtained from the tangential component of velocity during induction, figure 3.2.

For pre-mixed combustion the effect of turbulence is to break up, or wrinkle,
the flame front. There can be pockets of burnt gas in the unburnt gas and vice
versa. This increases the flame front area and speeds up combustion. Figure 3.3
shows a comparison of laminar and turbulent flame fronts.

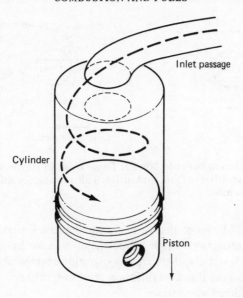

Figure 3.2   Swirl generation from tangential inlet passage

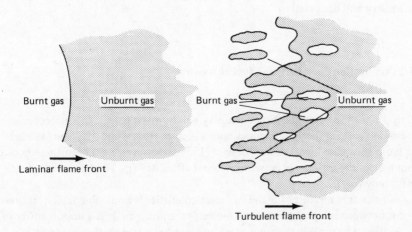

Figure 3.3   Comparison between laminar and turbulent flame fronts for pre-mixed
             combustion

For diffusion-controlled combustion the turbulence again enhances the burning velocity. Fuel is injected as a fine spray into air which is hot enough to
vaporise and ignite the fuel. The ordered air motion is also important because it
sweeps away the vaporised fuel and combustion products from the fuel droplets;
this is shown in figure 3.4.

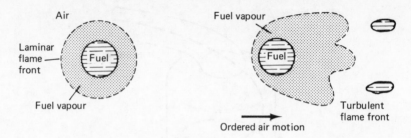

Figure 3.4   Comparison between laminar flame front in stagnant air with turbulent flame front and ordered air motion for diffusion-controlled combustion

Finally, it should be noted that there are compression ignition engines with pre-mixed combustion, for example 'model diesel' engines that use a carburetted ether-based fuel. Conversely, there are spark ignition engines such as some stratified charge engines that have diffusion processes; neither of these exceptions will be considered separately.

Before discussing combustion in spark ignition engines and compression ignition engines in any greater detail, it is necessary to study combustion chemistry and dissociation.

3.2 Combustion chemistry and fuel chemistry

Only an outline of combustion chemistry will be given here as the subject is treated in general thermodynamics books such as Rogers and Mayhew (1980a) or more specialised works like Goodger (1979). However, neither of these books emphasises the difference between rational efficiency (η_R) and aribtrary overall efficiency (η_O).

For reacting mixtures the use of molar quantities is invaluable since reactions occur between integral numbers of molecules, and the mole is a unit quantity of molecules. The mole is the amount of substance in a system that contains as many elementary entities as there are atoms in 0.012 kg of carbon 12. The normal SI unit is the kilomole (kmol), and the molar number (Avogadro constant) is 6.023×10^{26} entities/kmol.

Consider the reaction between two molecules of carbon monoxide (CO) and one molecule of oxygen (O_2) to produce two molecules of carbon dioxide (CO_2):

$$2CO + O_2 \rightarrow 2CO_2$$

There is conservation of mass, and conservation of the number of atoms.

It is often convenient to consider a unit quantity of fuel, for instance a kilomole, so the above reaction can be written in terms of kilomoles as

$$CO + \tfrac{1}{2}O_2 \rightarrow CO_2$$

A further advantage of using kilomoles is that, for the same conditions of temperature and pressure, equal volumes of gas contain the same number of moles. Thus the volumetric composition of a gas mixture is also the molar composition of the gas mixture. This is obviously not the case if liquid or solids are also present.

As well as the forward reaction of carbon monoxide oxidising to carbon dioxide, there will be a reverse reaction of carbon dioxide dissociating to carbon monoxide and oxygen:

$$CO + \tfrac{1}{2}O_2 \rightleftharpoons CO_2$$

When equilibrium is attained the mixture will contain all possible species from the reaction. In addition the oxygen can also dissociate

$$O_2 \rightleftharpoons 2O$$

With internal combustion engines, dissociation is important. Furthermore, there is not always sufficient time for equilibrium to be attained. Initially, complete combustion will be assumed.

Fuels are usually burnt with air, which has the following composition:

Molar   21.0 per cent O_2 79 per cent N_2^*
Gravimetric 23.2 per cent O_2 76.8 per cent N_2^*

Atmospheric nitrogen, N_2^*, represents all the constituents of air except oxygen and water vapour. Its thermodynamic properties are usually taken to be those of pure nitrogen. The molar masses (that is, the masses containing a kilomole of molecules) for these substances are:

$$O_2\ 31.999\ kg/kmol;\ N_2\ 28.013\ kg/kmol$$

$$N_2^*\ 28.15\ kg/kmol;\ air\ 28.96\ kg/kmol$$

When carbon monoxide is burnt with air the reaction (in kmols) is

$$CO + \tfrac{1}{2}\left(O_2 + \frac{79}{21}N_2^*\right) \rightarrow CO_2 + \tfrac{1}{2}\frac{79}{21}N_2^*$$

The nitrogen must be kept in the equation even though it does not take part in the reaction; it affects the volumetric composition of the products and the combustion temperature. The molar or volumetric air/fuel ratio is

$$1 : \tfrac{1}{2}\left(1 + \frac{79}{21}\right)$$

or $1 : 2.38$

The gravimetric air/fuel ratio is found by multiplying the number of moles by the respective molar masses — $(12 + 16)$ kg/kmol for carbon monoxide, and 29 kg/kmol for air:

$$1 . (12 + 16) : 2.38 (29)$$

or $1 : 2.47$

So far the reacting mixtures have been assumed to be chemically correct for complete combustion (that is, stoichiometric). In general, reacting mixtures are non-stoichiometric; these mixtures can be defined in terms of the excess air, theoretical air or equivalence ratio. Consider the same reaction as before with 25 per cent excess air, or 125 per cent theoretical air:

$$CO + \frac{1.25}{2} \left(O_2 + \frac{79}{21} N_2^* \right) \rightarrow CO_2 + \frac{0.25}{2} O_2 + \frac{1.25}{2} \cdot \frac{79}{21} N_2^*$$

The equivalence ratio

$$\phi = \frac{\text{stoichiometric air/fuel ratio}}{\text{actual air/fuel ratio}} = \frac{1}{1.25} = 0.8$$

The air/fuel ratio can be either gravimetric or molar; the usual form is gravimetric and this is often implicit.

Fuels are often mixtures of hydrocarbons, with bonds between carbon atoms, and between hydrogen and carbon atoms. During combustion these bonds are broken, and new bonds are formed with oxygen atoms, accompanied by a release of chemical energy. The principal products are carbon dioxide and water.

As combustion does not pass through a succession of equilibrium states it is irreversible, and the equilibrium position will be such that entropy is a maximum. The different compounds in fuels are classified according to the number of carbon atoms in the molecules. The size and geometry of the molecule have a profound effect on the chemical properties. Each carbon atom requires four bonds; these can be single bonds or combinations of single, double and triple bonds. Hydrogen atoms require a single bond.

An important family of compounds in petroleum (that is, petrol or diesel fuel) are the alkanes, formerly called the paraffins. Table 3.1 lists some of the alkanes; the different prefixes indicate the number of carbon atoms.

The alkanes have a general formula $C_n H_{2n+2}$, where n is the number of carbon atoms. Inspection shows that all the carbon bonds are single bonds, so the alkanes are termed 'saturated'. For example, propane has the structural formula

$$
\begin{array}{ccccccc}
 & H & & H & & H & \\
 & | & & | & & | & \\
H- & C & - & C & - & C & -H \\
 & | & & | & & | & \\
 & H & & H & & H &
\end{array}
$$

Table 3.1 Alkane family of compounds

Formula	Name	Comments	
CH_4	methane	'Natural gas' ⎫	LPG
C_2H_6	ethane	⎬	Liquefied
C_3H_8	propane ⎱	'Bottled gas' ⎭	Petroleum
C_4H_{10}	butane ⎰		Gases
C_5H_{12}	pentane ⎫		
C_6H_{14}	hexane	Liquids at room	
C_7H_{16}	heptane	temperature	
C_8H_{18}	octane ⎬		
.			
.			
.			
$C_{16}H_{34}$	cetane ⎭		
.			
.			
etc.			

When four or more carbon atoms are in a chain molecule it is possible to form isomers. Isomers have the same chemical formula but different structures, which often leads to very different chemical properties. Iso-octane is of particular significance for spark ignition engines; although it should be called 2,2,4-trimethyl-pentane, the isomer implied in petroleum technology is

$$
\begin{array}{ccccc}
 & & | & & \\
 & & -C- & & \\
 & | & | & | & | \\
-C- & C- & C- & C- & C- \\
 & | & | & | & | \\
 & -C- & & -C- & \\
 & | & & | &
\end{array}
\qquad \text{(Hydrogen atoms not shown)}
$$

Compounds that have straight chains with a single double bond are termed alkenes (formerly olefines); the general formula is C_nH_{2n}. An example is propylene, C_3H_6:

$$
\begin{array}{ccc}
 & H & H \\
 & | & \diagdown \\
H-C-C=C & \\
 & | \quad | & \diagup \\
 & H \quad H & H
\end{array}
$$

Such compounds are termed 'unsaturated' as the double bond can be split and extra hydrogen atoms added, a process termed 'hydrogenation'.

Most of the alkene content in fuels comes from catalytic cracking. In this process the less volatile alkanes are passed under pressure through catalysts such as silica or alumina at about $500°C$. The large molecule are decomposed, or cracked, to form smaller more volatile molecules. A hypothetical example might be

$$C_{20}H_{42} \rightarrow C_4H_8 + C_5H_{10} + C_4H_{10} + C_6H_{14} + C$$

alkane   alkenes   alkanes   carbon

A disadvantage of alkenes is that they can oxidise when the fuel is stored in contact with air. The oxidation products reduce the quality of the fuel and leave gum deposits in the engine. Oxidation can be inhibited by the addition of alkyl phenol, typically 100 ppm (parts per million by weight).

Hydrocarbons can also form ring structures, which can be saturated or unsaturated. Cyclo-alkanes are saturated and have a general formula C_nH_{2n}; in petroleum technology they are called naphthenes. An example is cyclopropane.

Aromatic compounds are unsaturated and based on the benzene molecule, C_6H_6. This has an unsaturated ring best represented as

The inner circle signifies two groups of three electrons, which form molecular bonds each side of the carbon atom plane. The structure accounts for the distinct properties of the aromatic compounds. Benzene and its derivatives occur in many crude oils but in particular they come from the distillation of coal.

The final class of fuels that have significance for internal combustion engines are the alcohols, in particular methanol (CH_3OH) and ethanol (C_2H_5OH):

The resurgent interest in alcohols is due to their manufacture from renewable energy sources, such as the destructive distillation of wood and fermentation of sugar respectively.

The calculations for the combustion of fuel mixtures are not complicated, but are best shown by worked solutions; see Examples 3.1 and 3.2. At this stage it is sufficient to note that the mean composition of molecules will be close to C_xH_{2x} and that most of the bonds will be single carbon–carbon or carbon–hydrogen. This implies that the stoichiometric (gravimetric) air/fuel ratio is always close to 14.8:1 (see problem 3.1). Furthermore, because of the similarity in bond structure, the calorific value and the density will vary only slightly for a range of fuels; this is shown in figure 3.5 using data from Blackmore and Thomas (1977) and Taylor (1966).

3.3 Combustion thermodynamics

Only combustion under the simple conditions of constant volume or constant pressure will be considered here. The gases, both reactants and products, will be considered as ideal (a gas that obeys the equation of state $pV = RT$, but has specific heat capacities that vary with temperature but not pressure). The molar enthalpy and molar internal energy of gases can be tabulated, Rogers and Mayhew (1980b), but if only molar enthalpy is tabulated, Haywood (1972), molar internal energy has to be deduced:

$$H \equiv U + pV \quad \text{or} \quad H \equiv U + R_0 T$$

The datum temperatures for these properties are arbitrary, but 25°C is convenient as it is the reference temperature for enthalpies of reaction.

Consider a rigid vessel containing a stoichiometric mixture of carbon monoxide and oxygen; the reactants (R) react completely to form carbon dioxide, the products (P). The first case, figure 3.6a, is when the container is insulated, so the process is adiabatic.

Apply the 1st Law of Thermodynamics:

$$\Delta E = Q - W$$

but $\quad Q = 0$ since the process is adiabatic
and $\quad W = 0$ since the container is rigid.
Thus $\quad \Delta E = \Delta U = 0$ since there is no change in potential or kinetic energy
and $\quad U_R = U_P$.

This process is represented by the line 1–2 on figure 3.6c, a plot that shows how the internal energy of the reactants and products vary with temperature. Temperature T_2 is the adiabatic combustion temperature.

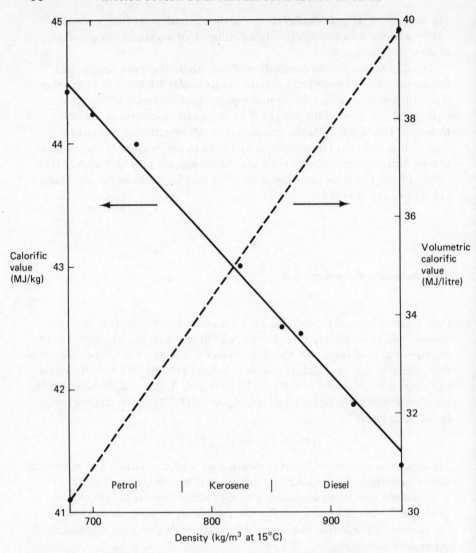

Figure 3.5 Variation in calorific value for different fuel mixtures (with
acknowledgement to Blackmore and Thomas (1977))

The second case to consider, figure 3.6b, is when the container is contrived
to be isothermal, perhaps by using a water bath. Again apply the 1st Law of
Thermodynamics:

$$\Delta E = Q - W$$

Again $W = 0$ since the container is rigid
but $Q \neq 0$ since the process is isothermal.

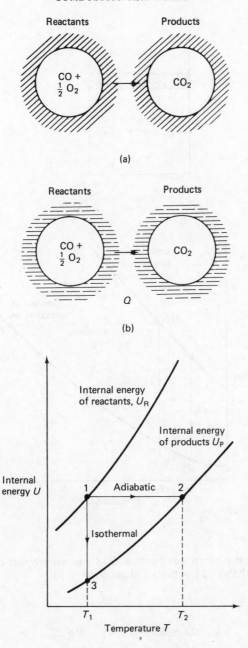

Figure 3.6 Constant-volume combustion. (a) Rigid insulated bomb calorimeter;
(b) rigid isothermal bomb calorimeter; (c) internal energy of
products and reactants

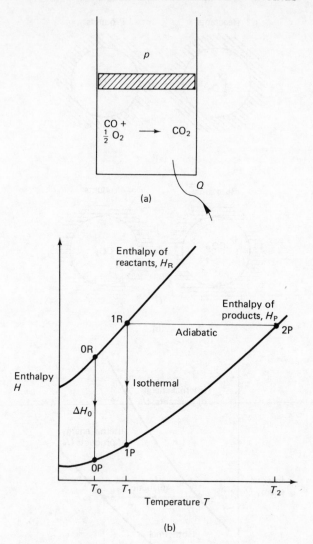

Figure 3.7   Constant-pressure combustion. (a) Constant-pressure calorimeter;
(b) enthalpy of reactants and products

Thus

$$\Delta E = \Delta U = Q = U_P - U_R \qquad (3.1)$$

Normally energy is released in a chemical reaction and Q is negative (exothermic reaction); if Q is positive the reaction is said to be endothermic. This method is used to find the constant volume or isochoric calorific value of a fuel in a combustion bomb.

Isochoric calorific value $= (CV)_{T,V} = -(\Delta U)_{T,V} = (U_R - U_P)_{T,V}$. The process is shown by the line 1–3 in figure 3.6c.

Combustion at constant pressure can be analysed by considering a rigid cylinder closed by a free-moving perfectly sealed piston, figure 3.7a. As before, the combustion can be contrived to be adiabatic or isothermal.

Again apply the 1st Law of Thermodynamics:

$$Q = \Delta U + W$$

In this case W is the displacement work. For constant pressure:

$$W = p\,\Delta V = \Delta p\,V$$

Thus

$$Q = \Delta U + \Delta p\,V = \Delta H = H_P - H_R \tag{3.2}$$

Adiabatic and isothermal combustion processes are shown on figure 3.7b. Calorific value normally refers to the isobaric isothermal calorific value, $CV = -\Delta H_{Tp} = (H_R - H_P)_{T,p}$. The datum temperature T_0 for tabulating calorific value is usually 298.15 K (25°C); this is denoted by $-\Delta H_0$. To evaluate calorific values at other temperatures, say T_1, recall that the calorific value at this temperature, $-\Delta H_{T_1 p}$, will be independent of the path taken.

Referring to figure 3.7b, 1R–0R–0P–1P will be equivalent to 1R–1P.

Thus

$$\Delta H_{T_1 p} = (H_{0P} - H_{1P}) + \Delta H_{T_0 p} + (H_{1R} - H_{0R}) \tag{3.3}$$

This principle is demonstrated by example 3.3. These calorific values can be found using steady-flow combustion calorimeters.

Since most combustion occurs at constant pressure (boilers, furnaces and gas turbines) isobaric calorific value $(-\Delta H_0)$ is tabulated instead of isochoric value $(-\Delta U_0)$; consequently there is a need to be able to convert from isobaric to isochoric calorific value.

Consider the reactants at the same temperature, pressure and volume (T, p, V_R). The difference between the calorific values is given by

$$(CV)_{p,T} - (CV)_{V,T} = (H_R - H_P)_{T,p} - (U_R - U_P)_{T,V_R}$$

Recalling that $H \equiv U + pV$

$$(CV)_{p,T} - (CV)_{V,T} = (U_R - U_P)_{T,p} + p(V_R - V_P)_{T,p} - (U_R - U_P)_{T,V_R}$$

and assuming that internal energy is a function only of temperature and not of pressure, that is

$$(U_P)_{p,T} = (U_P)_{V_R,T}$$

thus

$$(CV)_{p,T} - (CV)_{V,T} = p\,[V_R - V_P]_{p,T}$$

Neglect the volumes of any liquids or solids, and assume ideal gas behaviour, $pV = nR_0T$; so

$$(CV)_{p,T} - (CV)_{V,T} = (n_{GR} - n_{GP})R_0T \qquad (3.4)$$

where n_{GR} number of moles of gaseous reactants/unit of fuel
n_{GP} number of moles of gaseous products/unit of fuel.

The difference between the calorific values is usually very small.

In most combustion problems any water produced by the reaction will be in the vapour state. If the water were condensed, the calorific value would be increased and then be referred to as the higher calorific value (HCV). The relationship between higher and lower calorific value (LCV) is

$$HCV - LCV = yh_{fg} \qquad (3.5)$$

where y is the mass of water per unit quantity of fuel and h_{fg} is the enthalpy of evaporation of water at the temperature under consideration.

For hydrocarbon fuels the difference is significant, but lower calorific value is invariably used or implied. See example 3.3 for the use of calorific values. Similarly the state of the fuel must be specified, particularly if it could be liquid or gas. However, the enthalpy of vaporisation for fuels is usually small compared to their calorific value. For example, for octane at 298.15 K (Rogers and Mayhew (1980b)):

$$H_{fg} = 41\,500 \text{ kJ/kmol}$$
$$\Delta H_0 = -5\,116\,180 \text{ kJ/kmol}$$

3.4 Dissociation

Dissociation has already been introduced in this chapter by discussing the dissociation of carbon dioxide

$$CO + \tfrac{1}{2}O_2 \rightleftharpoons CO_2$$

According to Le Châtelier's Principle, an equilibrium will always be displaced in such a way as to minimise any changes imposed from outside the system. This equilibrium can be affected in three ways: a change in concentration of a constituent, a change in system pressure, or a change in temperature. Considering one change at a time

(i) Suppose excess oxygen were added to the system.
     The reaction would move in the forwards direction, as this would reduce the concentration of oxygen.

(ii)   Suppose the system pressure were increased.
       Again the reaction would move in the forwards direction, as this reduces
       the total number of moles (n), and the pressure reduces since

$$p V = n R_0 T$$

(iii) Suppose the temperature of the system were raised.
       The reaction will move in a direction that absorbs heat; for this particular
       reaction, that will be the reverse direction.

Care must be taken in defining the forward and reverse directions as these are
not always self-evident. For example, take the water gas reaction

$$CO_2 + H_2 \rightleftharpoons CO + H_2O$$

To study this matter more rigorously it is necessary to introduce the concept of
equilibrium constants, K; these are also called dissociation constants.
   Consider a general reaction of

$$a \text{ kmols } A + b \text{ kmols } B \rightleftharpoons c \text{ kmols } C + d \text{ kmols } D$$

It can be shown by use of a hypothetical device, the Van t'Hoff equilibrium box,
that

$$K = \frac{(p'_c)^c (p'_d)^d}{(p'_a)^a (p'_b)^b} \tag{3.6}$$

The derivation can be found in thermodynamics texts such as Rogers and
Mayhew (1980a). K is a function only of temperature and will have units of
pressure to the power $(c + d - a - b)$. p', the partial pressure of a component
is defined as

$$p'_a = \frac{a}{a + b + c + d} \, p, \text{ where } p \text{ is the system pressure}$$

To solve problems involving dissociation introduce a variable x

$$CO + \tfrac{1}{2}O_2 \rightarrow (1 - x)CO_2 + xCO + \frac{x}{2}O_2$$

$$K = \frac{p'_{CO_2}}{(p'_{CO})(p'_{O_2})^{\frac{1}{2}}} \text{ where } p'_{CO_2} = \frac{1 - x}{(1 - x) + x + \dfrac{x}{2}} \, p, \text{ etc.} \tag{3.7}$$

so

$$K = \frac{1 - x}{x \, (x/2)^{\frac{1}{2}}} \left(\frac{1 + \dfrac{x}{2}}{p} \right)^{\frac{1}{2}}$$

As the equilibrium constant varies strongly with temperature, it is most convenient to tabulate $\log_{10}(K)$; the numerical value will also depend on the pressure

units adopted (Rogers and Mayhew (1980b) or Haywood (1972)). Again, the use of equilibrium constants is best shown by a worked problem, example 3.5. With internal combustion engines there will be several dissociation mechanisms occurring, and the simultaneous solution is best performed by computer.

3.5 Combustion in spark ignition engines

Combustion either occurs normally – with ignition from a spark and the flame front propagating steadily throughout the mixture – or abnormally. Abnormal combustion can take several forms, principally pre-ignition and self-ignition. Pre-ignition is when the fuel is ignited by a hot spot, such as the exhaust valve or incandescent carbon combustion deposits. Self-ignition is when the pressure and temperature of the fuel/air mixture are such that the remaining unburnt gas ignites spontaneously. Pre-ignition can lead to self-ignition and vice versa; these processes will be discussed in more detail after normal combustion has been considered.

3.5.1 Normal combustion

When the piston approaches the end of the compression stroke, a spark is discharged between the sparking plug electrodes. The spark leaves a small nucleus of flame that propagates into the unburnt gas. Until the nucleus is of the same order of size as the turbulence scale, the flame propagation cannot be enhanced by the turbulence. This causes a delay period of approximately constant duration. Figure 3.8 compares the indicator diagrams for the cases where a mixture is ignited, and where it is not ignited. The point at which the pressure traces diverge is ill-defined, but is used to denote the end of the delay period. The delay period is typically of about 0.5 ms duration, which corresponds to about $7\frac{1}{2}°$ of crank angle at 2500 rpm. The delay period depends on the temperature, pressure and composition of the fuel/air mixture, but it is a minimum for slightly richer than stoichiometric mixtures.

The end of the second stage of combustion is also ill-defined on the pressure diagram, but occurs shortly after the peak pressure. The second stage of combustion is affected in the same way as delay period, and also by the turbulence. This is very fortunate since turbulence increases as the engine speed increases, and the time for the second stage of combustion reduces almost in proportion. In other words, the second stage of combustion occupies an approximately constant number of crank angle degrees. In practice, the maximum cylinder pressure usually occurs 5–20° after top dead centre (Benson and Whitehouse

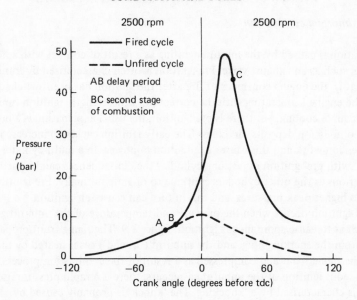

Figure 3.8 Hypothetical indicator diagram for a spark ignition engine

(1979)). It is normal for combustion to be complete before the exhaust valve is opened.

Since combustion takes a finite time, the mixture is ignited before top dead centre (btdc), at the end of the compression stroke. This means that there is a pressure rise associated with combustion before the end of the compression stroke, and an increase in the compression (negative) work. Advancing the ignition timing causes both the pressure to rise before top dead centre and also the compression work to increase. In addition the higher pressure at top dead centre leads to higher pressures during the expansion stroke, and to an increase in the expansion (positive) work. Obviously there is a trade-off between these two effects, and this leads to an optimum ignition timing. Since the maximum is fairly insensitive to ignition timing, the minimum ignition advance is used; this is referred to as 'minimum (ignition) advance for best torque' (MBT). By using the minimum advance, the peak pressures and temperatures in the cylinder are reduced; this helps to restrict heat transfer, engine noise, emissions, and susceptibility to abnormal combustion. Similar arguments apply to compression ignition engines.

During the early stages of combustion, while the flame nucleus is still small, it can be displaced from the sparking plug region by large-scale flows in the cylinder. This can occur in a random way, and can have a significant effect on the subsequent propagation of combustion. This is readily shown by the non-repeatability of consecutive indicator diagrams from an engine, and is called variously 'cyclic dispersion' or 'cyclic variation'.

3.5.2 Abnormal combustion

Pre-ignition is caused by the mixture igniting as a result of contact with a hot surface, such as an exhaust valve. Pre-ignition is often characterised by 'running-on ; that is, the engine continues to fire after the ignition has been switched off.

If the engine is operating with the correct mixture strength, ignition timing and adequate cooling, yet there is pre-ignition, the usual explanation is a build-up of combustion deposits, or 'coke'. The early ignition causes an increase in the compression work and this causes a reduction in power. In a multi-cylinder engine, with pre-ignition in just one cylinder, the consequences can be particu-larly serious as the other cylinders continue to operate normally. Pre-ignition leads to higher peak pressures, and this in turn can cause self-ignition.

Self-ignition occurs when the pressure and temperature of the unburnt gas are such as to cause spontaneous ignition, figure 3.9. The flame front propagates away from the sparking plug, and the unburnt (or 'end') gas is heated by radiation from the flame front and compressed as a result of the combustion process. If spontaneous ignition of the unburnt gas occurs, there is a rapid pressure rise which is characterised by a 'knocking' The 'knock' is probably caused by resonances of the combustion chamber contents. As a result of knocking, the thermal boundary layer at the combustion chamber walls can be destroyed. This causes increased heat transfer which might then lead to certain surfaces causing pre-ignition.

In chapter 2 it was shown how increasing the compression ratio should improve engine performance. Unfortunately, raising the compression ratio also increases the susceptibility to knocking. For this reason, much research has centred on the fundamental processes occurring with knock. These mechanisms are discussed in section 3.7.

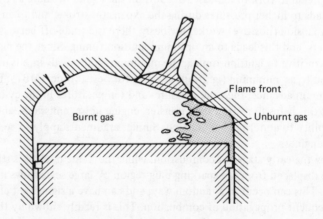

Figure 3.9   Combustion in a spark ignition engine

3.6 Combustion in compression ignition engines

Near the end of the compression stroke, liquid fuel is injected as one or more jets. The injector receives fuel at very high pressures in order to produce rapid injection, with high velocity jet(s) of small cross-sectional area; in all but the largest engines there is a single injector. The fuel jets entrain air and break up into droplets; this provides rapid mixing which is essential if the combustion is to occur sufficiently fast. Sometimes the fuel jet is designed to impinge on to the combustion chamber wall; this can help to vaporise the fuel and break up the jet. There will be large variations in fuel/air mixtures on both a large and small scale within the combustion chamber. Figure 3.10 shows the indicator diagram for a compression ignition engine; when compared to figure 3.8 (spark ignition engine) it can be seen immediately that the pressures are higher, especially for the unfired cycle. Referring to figure 3.10, there are several stages of combustion, not distinctly separated:

(i) Ignition delay, AB. After injection there is initially no apparent deviation from the unfired cycle. During this period the fuel is breaking up into droplets being vaporised, and mixing with air. Chemical reactions will be starting, albeit slowly.

(ii) Rapid or uncontrolled combustion, BC. A very rapid rise in pressure caused by ignition of the fuel/air mixture prepared during the ignition delay period.

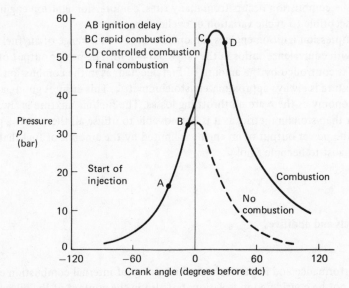

Figure 3.10   Hypothetical indicator diagram for a compression ignition engine

(iii)   Controlled combustion, CD. Combustion occurs at a rate determined by the preparation of fresh air/fuel mixture.

(iv)   Final combustion D. As with controlled combustion the rate of combustion is governed by diffusion until all the fuel or air is utilised.

As with spark ignition engines the initial period is independent of speed, while the subsequent combustion occupies an approximately constant number of crank angle degrees. In order to avoid too large a rapid combustion period, the initial fuel injection should be carefully controlled. The 'rapid' combustion period can produce the characteristic 'diesel knock'. Again this is caused by a sudden pressure rise, but is due to self-ignition occurring too slowly. Its cure is the exact opposite to that used in spark ignition engines; fuels in compression ignition engines should self-ignite readily. For a given fuel and engine, diesel knock can be reduced by avoiding injection of too much fuel too quickly. Some systems inject a small quantity of fuel before the main injection, a system known as pilot injection. Alternatively, the engine can be modified to operate with a higher compression ratio. This increases the temperature and pressure during the compression stroke, and this will reduce the ignition delay period.

To obtain maximum output the peak pressure should occur about 10–20° after top dead centre. Sometimes the injection is later, in order to retard and to reduce the peak pressure.

Combustion photography in compression ignition engines can be very useful, as it shows the effectiveness of the injection process. The fuel ignites spontaneously at many sites, and produces an intense flame. There is significant radiation from the flame front, and this is important for vaporising the fuel. Since the combustion occurs from many sites, compression ignition engines are not susceptible to cyclic variation or cyclic dispersion.

Compression ignition engines can operate over a wide range of air/fuel mixtures with equivalence ratios in the range 0.04–0.90. The power output of the engine is controlled by the amount of fuel injected, as in the combustion region the mixture is always approximately stoichiometric. This ensures good part load fuel economy as there are no throttling losses. The fuel/air mixture is always weaker than stoichiometric, as it is not possible to utilise all the air. At a given speed the power output of an engine is limited by the amount of fuel that causes the exhaust to become smoky.

3.7 Fuels and additives

The performance and in particular fuel economy of internal combustion engines should not be considered in isolation, but also in the context of the oil refinery. Oil refining and distribution currently has an overall efficiency of about 88 per cent (Francis and Woollacott (1981)). This efficiency would be changed if the demand for the current balance of products was changed. Oil refining can be compared to the work of a butcher; each process has a raw material (a barrel of oil or a carcass) that has to be cut in such a way that all the products can be sold at a competitive price. Just as there are different animals, there are also different types of crude oil, depending on the source. However, the oil refinery can change the type of product in additional ways, such as cracking, although there is an energy cost associated with this.

The energy content of a typical petrol is 44 MJ/kg or 31.8 MJ/litre and of a typical diesel fuel 42 MJ/kg or 38.15 MJ/litre; associated with these is an energy content at the refinery of typically 2.7 MJ/kg and 1.65 MJ/kg (Francis and Woollacott (1981)). This gives an effective primary energy of 46.7 MJ/kg or 35.5 MJ/litre for petrol and 43.65 MJ/kg or 39.95 MJ/litre for diesel fuel. There may be circumstances in which it is more appropriate to use primary energy density than to use calorific value. These figures also highlight the difference in energy content of unit volume of fuel. This is often overlooked when comparing the fuel economy of vehicles on a volumetric basis.

3.7.1 Characteristics of petrol

The properties of petrol are discussed thoroughly by Blackmore and Thomas (1977). The two most important characteristics of petrol are its volatility and octane number (its resistance to self-ignition).

Volatility is expressed in terms of the volume percentage that is distilled at or below fixed temperatures. If a petrol is too volatile, when it is used at high ambient temperatures the petrol is liable to vaporise in the fuel lines and form vapour locks. This problem is most pronounced in vehicles that are being re-started, since under these conditions the engine compartment is hottest. If the fuel is not sufficiently volatile the engine will be difficult to start, especially at low ambient temperatures. The volatility also influences the cold start fuel economy. Spark ignition engines are started on very rich mixtures, and continue to operate on rich mixtures until they reach their normal operating temperature; this is to ensure adequate vaporisation of fuel. Increasing the volatility of the petrol at low temperatures will evidently improve the fuel economy during and after starting. Blackmore and Thomas (1977), point out that in the USA as much as 50 per cent of all petrol is consumed on trips of 10 miles or less. Short journeys have a profound effect on vehicle fuel economy, yet fuel consumption figures are invariably quoted for steady-state conditions.

Fuel volatility is specified in British Standard 4040: 1978, and these data are compared with typical fuel specifications from Blackmore and Thomas (1977) in table 3.2. This is plotted with further data in figure 3.11.

Table 3.2 shows how the specification of petrol varies to suit climatic conditions. Petrol stored for a long time in vented tanks is said to go stale; this refers

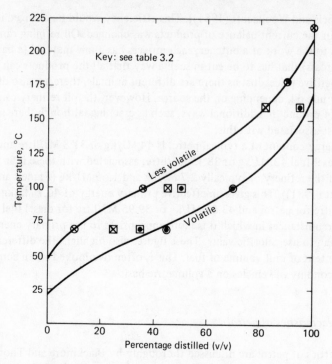

Figure 3.11   Distillation curves for petrol (with acknowledgement to
               Blackmore and Thomas (1977))

Table 3.2 Volatility of different petrol blends

	BS4040		Less volatile	Volatile	North-west Europe		Central Africa
	Min.	Max.			Summer	Winter	
Distillate evaporated at 70°C (per cent V/V)	10	45	10	42	25	35	10
Distillate evaporated at 100°C (per cent V/V)	36	70	38	70	45	50	38
Distillate evaporated at 160°C (per cent V/V)			80	98	80	95	80
Distillate evaporated at 180°C (per cent V/V)	90						
Final boiling point °C	–	220					
Residue (per cent V/V)		2					
Symbol used in figure 3.11	⊗	⊙	–	–	⊠	⊡	–

to the loss of the more volatile components that are necessary for easy engine starting.

The octane number of a fuel is a measure of its anti-knock performance. A scale of 0-100 is devised by assigning a value of 0 to n-heptane (a fuel prone to knock), and a value of 100 to iso-octane (a fuel resistant to knock). A 95 octane fuel has the performance equivalent to that of a mixture of 95 per cent iso-octane and 5 per cent n-heptane by volume. The octane requirement of an engine varies with compression ratio, geometrical and mechanical considerations, and also its operating conditions. There are two commonly used octane scales, research octane number (RON) and motor octane number (MON), covered by British Standards 2637: 1978 and 2638: 1978 respectively. Both standards refer to the *Annual Book of ASTM* (American Society for Testing and Materials) *Standards Part 47 — Test Methods for Rating Motor, Diesel and Aviation Fuels*.

The tests for determining octane number are performed using the ASTM–CFR (Cooperative Fuel Research) engine; this is a variable compression ratio engine similar to the Ricardo E6 engine. In a test the compression ratio of the engine is varied to obtain standard knock intensity. With the same compression ratio two reference fuel blends are found whose knock intensities bracket that of the sample. The octane rating of the sample can then be found by interpolation. The different test conditions for RON and MON are quoted in *ASTM Standards Part 47*, and are summarised in table 3.3.

Table 3.3 shows that the motor octane number has more severe test conditions since the mixture temperature is greater and the ignition occurs earlier. There is not necessarily any correlation between MON and RON as the way fuel components of different volatility contribute to the octane rating will vary. Furthermore, when a carburetted engine has a transient increase in load, excess fuel is supplied. Under these conditions it is the octane rating of the more volatile components that determines whether or not knock occurs. The minimum octane requirements for different grades of petrol are given by BS4040, see table 3.4. A worldwide summary of octane ratings is published by the Associated Octel Co. Ltd, London.

The attraction of high octane fuels is that they enable high compression ratios to be used. Higher compression ratios give increased power output and improved economy. This is shown in figure 3.12 using data from Blackmore and Thomas (1977). The octane number requirements for a given compression ratio vary widely, but typically a compression ratio of 7.5 requires 85 octane fuel, while a compression ratio of 10.0 would require 100 octane fuel. There are even wide variations in octane number requirements between supposedly identical engines.

Of the various fuel additives, those that increase octane numbers have greatest significance. In 1922 Midgely and Boyd discovered that lead-based compounds improved the octane rating of fuels. By adding 0.5 grams of lead per litre, the octane rating of the fuel is increased by about 5 units. The lead additives take the form of lead alkyls, either tetramethyl lead $(CH_3)_4Pb$, or tetraethyl lead $(C_2H_5)_4Pb$.

Table 3.3 Summary of RON and MON test conditions

Test conditions	Research octane number	Motor octane number
Engine speed, rpm	600 ± 6	900 ± 9
Crankcase oil, SAE grade	30	30
Oil pressure at operating		
temperature, psi	25-30	25-30
Crankcase oil temperature	$135 \pm 15°F$ ($57 \pm 8.5°C$)	$135 \pm 15°F$ ($57 \pm 8.5°C$)
Coolant temperature		
Range	$212 \pm 3°F$ ($100 \pm 1.5°C$)	$212 \pm 3°F$ ($100 \pm 1.5°C$)
Constant within	$\pm 1°F$ ($0.5°C$)	$\pm 1°F$ ($0.5°C$)
Intake air humidity, grains		
of water per lb. of dry air	25-50	25-50
Intake air temperature	See *ASTM Standard Part 47*	$100 \pm 5°F$ ($38 \pm 2.8°C$)
Mixture temperature	−	$300 \pm 2°F$ ($149 \pm 1.1°C$)
Spark advance, deg. btdc	13	14-26 depending on compression ratio
Spark plug gap, in.	0.020 ± 0.005	0.020 ± 0.005
Breaker point, gap, in.	0.020	0.020
Valve clearances, in.		
Intake	0.008	0.008
Exhaust	0.008	0.008
Fuel/air ratio	Adjusted for maximum knock	

Table 3.4 Octane number requirements for different fuel grades

Grade designation	RON	MON
5 star	100.0	86.0
4 star	97.0	86.0
3 star	94.0	82.0
2 star	90.0	80.0

Since the active ingredient is lead, the concentration of the additives is expressed in terms of the lead content. Thus

$$0.5 \text{ g Pb/l} = 0.645 \text{ g } (CH_3)_4 Pb/l$$

and

$$0.5 \text{ g Pb/l} = 0.780 \text{ g } (C_2H_5)_4 Pb/l$$

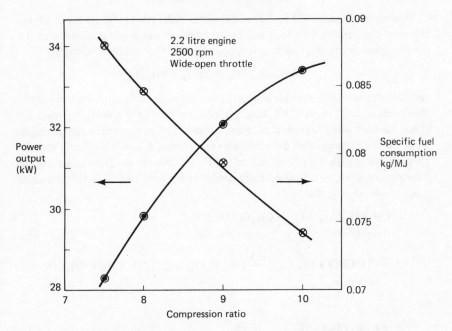

Figure 3.12 Effect of changing compression ratio on engine power output and fuel economy (with acknowledgement to Blackmore and Thomas (1977))

Most countries now have restrictions on the use of lead in fuels for environmental reasons. As well as the possible dangers of lead pollution, catalysts for the conversion of other engine pollutants are made inactive by lead. However, manufacturers of lead additives claim that suitable filters could be installed in exhaust systems to remove the lead particulates.

To understand how lead alkyls can inhibit knocking, the chemical mechanism involved in knocking must be considered in more detail. Two possible causes of knocking are cool flames or low-temperature auto-ignition, and high-temperature auto-ignition. Cool flames can occur in many hydrocarbon fuels and are studied by experiments with fuel/oxygen mixtures in heated vessels. If some mixtures are left for a sufficient time, a flame is observed at temperatures that are below those for normal self-ignition; the flames are characterised by the presence of peroxide and aldehyde species. Engine experiments have been conducted in which the concentrations of peroxide and aldehyde species have been measured, giving results that imply the presence of a cool flame. However, knock was obtained only with higher compression ratios, which implied that it is a subsequent high-temperature auto-ignition that causes the rapid pressure rise and knock. Cool flames have not been observed with methane and benzene, so when knock occurs with these fuels it is a single-stage high-temperature auto-ignition effect.

Downs and Wheeler (1951–52) and Downs *et al.* (1961) discuss the possible chemical mechanisms of knock and how tetraethyl lead might inhibit knock. In the combustion of, say, heptane, it is unlikely that the complete reaction

$$C_7H_{16} + 11O_2 \rightarrow 7CO_2 + 8H_2O$$

can occur in one step. A gradual degradation through collisions with oxygen molecules is much more likely, finally ending up with CO_2 and H_2O. This is a chain reaction where oxygenated hydrocarbons such as aldehydes and peroxides will be among the possible intermediate compounds. A possible scheme for propane starts with the propyl radical $(C_3H_7)^-$, involves peroxide and aldehyde intermediate compounds, and finally produces a propyl radical so that the chain reaction can then repeat:

$$CH_3CH^-CH_3 + O_2 \rightarrow CH_3CH(OO)CH_3$$
propyl radical   hydroperoxide

$$\rightarrow CH_3CHO + CH_3O^- \xrightarrow{+C_3H_8} CH_3OH + CH_3CH^-CH_3 + CH_3CHO$$
     aldehyde   alcohol

or

$$\rightarrow C_2H_5CHO + OH^-$$
               hydroxyl
               radical

There are many similar chain reactions that can occur and some of the possibilities are discussed by Lewis and von Elbe (1961) in greater detail.

Tetraethyl lead improves the octane rating of the fuel by modifying the chain reactions. During the compression stroke, the lead alkyl decomposes and reacts with oxygenated intermediary compounds to form lead oxide, thereby combining with radicals that might otherwise cause knock.

The suggested mechanism is as follows:

$$PbO + OH \rightarrow PbO(OH)$$
$$PbO(OH) + OH \rightarrow PbO_2 + H_2O$$
$$PbO_2 + R \;\;\; \rightarrow PbO + RO$$

where R is a radical such as the propyl radical, $C_3H_7^-$.

One disadvantage with the lead alkyls is that lead compounds are deposited in the combustion chamber. These can be converted to more volatile lead halides by alkyl halide additives such as dichloroethane, $(C_2Cl_2H_4)$ or dibromoethane $(C_2Br_2H_4)$. However, some lead halides remain in the combustion chamber and these deposits can impair the insulation of spark plugs, and thus lead to misfiring. By adding aryl phosphates to petrol, lead halides are converted to phosphates, which have greater electrical resistivity. A second benefit is that lead phosphates are less prone to cause pre-ignition by surface ignition. These and other additives are discussed more fully by Blackmore and Thomas (1977).

Alcohols have certain advantages as fuels, particularly in countries without oil resources, or where there are sources of the renewable raw materials for producing methanol (CH_3OH) or ethanol (C_2H_5OH). Car manufacturers have extensive programmes for developing alcohol-fuelled vehicles (Ford (1982)). Alcohols can also be blended with oil-derived fuels and this improves the octane ratings. Both alcohols have high octane ratings (ethanol has a RON of 106) and high enthalpy of vaporisation; this improves the volumetric efficiency but can cause starting problems. For cold ambient conditions it may be necessary to start engines with petrol. The other main disadvantages are the lower energy densities (about half that of petrol for methanol and two-thirds for ethanol), and the miscibility with water.

3.7.2 Characteristics of diesel fuel

The most important characteristic of diesel fuel is the cetane number, as this indicates how readily the fuel self-ignites. Viscosity is also important, especially for the lower-grade fuels used in the larger engines; sometimes it is necessary to have heated fuel lines. Another problem with diesel fuels is that, at low temperatures, the high molecular weight components can precipitate to form a waxy deposit. This is defined in terms of the cold filter plugging point.

The properties of different fuel oils are specified in BS2869: 1970. The two fuel specifications quoted in table 3.5 are for high-quality automotive diesel fuel (A1), and general-purpose diesel fuel (A2).

The flashpoint is the temperature to which the liquid has to be heated for the vapour to form a combustible mixture with air at atmospheric pressure. Since the flashpoint of diesel fuel is at least $55°C$, this makes it a safer fuel to store than either petrol or kerosene. The flashpoints of petrol and kerosene are about $-40°C$ and $30°C$ respectively.

If an engine runs on a fuel with too low a cetane number, there will be diesel knock. Diesel knock is caused by too rapid combustion and is the result of a long ignition delay period, since during this period fuel is injected and mixes with air to form a combustible mixture. Ignition occurs only after the pressure and temperature have been above certain limits for sufficient time, and fuels with high cetane numbers are those that self-ignite readily.

As with octane numbers, a scale of 0–100 is constructed by assigning a value of 0 to α-methylnaphthalene ($C_{10}H_7CH_3$, a naphthenic compound with poor self-ignition qualities), and a value of 100 to n-cetane ($C_{16}H_{34}$, a straight-chain alkane with good self-ignition qualities). A 65 cetane fuel would have ignition delay performance equivalent to that of a blend of 65 per cent n-cetane and 35 per cent α-methylnaphthalene by volume.

The tests for determining cetane number in BS5580: 1978 refer to the *Annual Book of ASTM Standards, Part 47*. The tests are performed with an ASTM–CFR engine equipped with a special instrument to measure ignition delay. With

Table 3.5 Specifications for diesel fuels

Property		BS 2869	
		Class A1	Class A2
Viscosity, kinematic at	min.	1.6	1.6
37.8°C (centistokes)	max.	6.0	6.0
Cetane number	min.	50	45
Carbon residue, Ramsbottom per cent by mass on 10 per cent residue	max.	0.2	0.2
Distillation, recovery at 357°C, per cent by volume	min.	90	90
Flashpoint, closed Pensky Martins °C	min.	55	55
Water content, per cent by volume	max.	0.05	0.05
Sediment, per cent	max.	0.01	0.01
Ash, per cent by mass	max.	0.01	0.01
Sulphur	max.	0.5	0.8
Copper corrosion test	max.	1	1
Cold filter plugging point (°C) max.	Summer	0 Mar./Nov.	0 Mar./Nov.
	Winter	−9 Dec./Feb.	−9 Oct./Feb.

standard operating conditions the compression ratio of the engine is adjusted to give a standard delay period with the fuel being tested. The process is repeated with reference fuel blends to find the compression ratios for the same delay period. When the compression ratio of the fuel being tested is bracketed by the reference fuels, the cetane number of the test fuel is found by interpolation.

As would be expected, fuels with high cetane numbers have low octane numbers and vice versa. This relationship is shown in figure 3.13 using data from Taylor (1968); surprisingly there is a single line, thus showing independence of fuel composition.

Additives in diesel fuel to improve the cetane number are referred to as ignition accelerators. Their concentrations are greater than those of anti-knock additives used in petrol. Typically an improvement of 6 on the cetane scale is obtained by adding 1 per cent by volume of amyl nitrate, $C_5H_{11}ONO_2$. Other effective substances are ethyl nitrate, $C_2H_5ONO_2$ and ethyl nitrite C_2H_5ONO.

Ignition delay is most pronounced at slow speeds because of the reduced temperature and pressure during compression. Cold-starting can be a problem, and is usually remedied by providing a facility on the injector pump to inject excess fuel. Under severe conditions, additional starting aids such as heaters

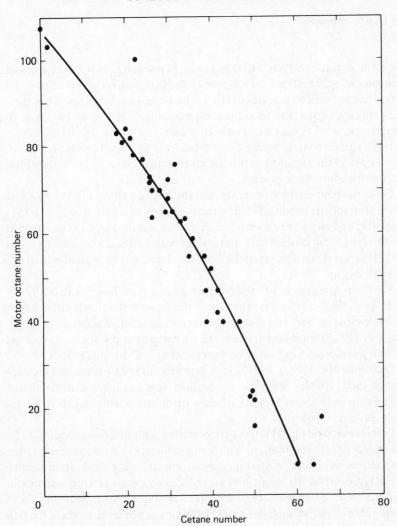

Figure 3.13   Relationship between cetane number and octane number for
                petroleum-derived fuels (adapted from Taylor (1968))

may be needed, or volatile fuels with high cetane numbers, such as ether, can be
added to the intake air.

Sometimes a cetane index is used, as the only information needed is fuel
viscosity and density with no need for engine tests. The cetane index can be
used only for straight petroleum distillates without additives. Other fuels that
are suitable for diesel engines are derived from coal and vegetable oils. Interest
in these alternative fuels will increase as the cost of petroleum rises and the
quality of petroleum-derived diesel fuel reduces.

3.8 Engine emissions

The term 'engine emissions' refers primarily to pollutants in the engine exhaust. Examples of pollutants are carbon monoxide (CO), various oxides of nitrogen (NO_x) and unburnt hydrocarbons (HC). These emissions are worse from the spark ignition engine than from the compression ignition engine. Emissions from compression ignition engines are primarily soot, and odour associated with certain hydrocarbons. Recently concern has been expressed about possible carcinogens in the exhaust but it is not clear if these come from the diesel fuel or from the combustion process.

Concern about emissions developed in the 1960s, particularly in Los Angeles where atmospheric conditions led to the formation of a photochemical smog from NO_x and HC. Exhaust emission legislation is historically and geographically too involved for discussion here, but is dealt with by Blackmore and Thomas (1977). Strictest controls are in the USA and Japan but European legislation is also building up.

The concentrations of CO and NO_x are greater than those predicted by equilibrium thermodynamics. The rate of the forward reaction is different from the backward reaction, and there is insufficient time for equilibrium to be attained. The chemical kinetics involved are complex and work is still proceeding to try and predict exhaust emissions (Mattavi and Amann (1980)).

Emissions of CO, NO_x and HC vary between different engines and are dependent on such variables as ignition timing, load, speed and, in particular, fuel/air ratio. Figure 3.14 shows typical variations of emissions with fuel/air ratio for a spark ignition engine.

Carbon monoxide (CO) is most concentrated with fuel-rich mixtures, as there will be incomplete combustion. With lean mixtures, CO is always present owing to dissociation, but the concentration reduces with reducing combustion temperatures. Hydrocarbon (HC) emissions are reduced by excess air (fuel-lean mixtures) until the reduced flammability of the mixtures causes a net increase in HC emissions. These emissions originate from the flame quench layer — where the flame is extinguished by cold boundaries; regions like piston ring grooves can be particularly important. The outer edge of the quench regions can also contribute to the CO and aldehyde emissions.

The formation of NO_x is more complex since it is dependent on a series of reactions such as the Zeldovich mechanism:

$$O_2 \rightleftharpoons 2O$$
$$O + N_2 \rightleftharpoons NO + N$$
$$N + O_2 \rightleftharpoons NO + O$$

Some of the modifications to this reaction and the effects of different operating conditions are discussed by Benson and Whitehouse (1979). Chemical kinetics show that the formation of NO and other oxides of nitrogen increase very

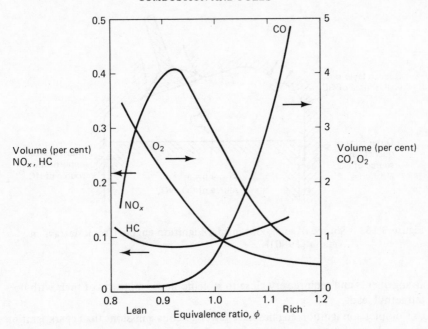

Figure 3.14   Spark ignition engine emissions for different fuel/air ratios
(courtesy of Johnson Matthey)

strongly with increasing flame temperature. This would imply that the highest
concentration of NO_x should be for slightly rich mixtures, those that have the
highest flame temperature. However, NO_x formation will also be influenced by
the flame speed. Lower flame speeds with lean mixtures provide a longer time for
NO_x to form. Similarly NO_x emissions increase with reduced engine speed. The
sources of different emissions are shown in figure 3.15.

Emissions of HC and CO can be reduced by operating with lean mixtures;
this has the disadvantage of reducing the engine power output. It is also difficult
to ensure uniform mixture distribution to each cylinder in multi-cylinder engines.
Alternatively, exhaust gas catalytic reactors or thermal reactors can complete
the oxidation process; if necessary extra air can be admitted.

The ways of reducing NO_x emissions are more varied. If either the flame
temperature or speed is reduced, the NO_x emissions will also be reduced.
Retarding the ignition is very effective as this reduces the peak pressure and
temperature, but it has an adverse effect on power output and economy. Another
approach is to increase the concentration of residuals in the cylinder by exhaust
gas recirculation (EGR). EGR lowers both flame temperature and speed, so giving
useful reductions in NO_x. Between 5 and 10 per cent EGR is likely to halve NO_x
emissions. However, EGR lowers the efficiency and reduces the lean combustion
limit. Catalysts can be used to reduce the NO_x to oxygen and nitrogen but this is
difficult to arrange if CO and HC are being oxidised. Such systems have complex

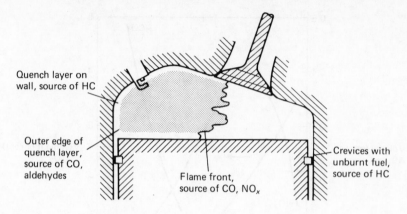

Figure 3.15   Source of emissions in spark ignition engine (from Mattavi and
Amann (1980))

arrangements and require very close to stoichiometric mixtures of fuels with no
tetraethyl lead.

Compression ignition engines have fewer gaseous emissions than spark ignition
engines, but compression ignition engines have greater particulate emissions. The
equivalence ratio in a diesel engine is always less than unity (fuel lean), and this
accounts for the low CO emissions, about 0.1 per cent by volume. Hydrocarbon
emission (unburnt fuel) is also less, but rises towards the emission level of spark
ignition engines as the engine load (bmep) rises.

The emissions of NO_x are about half those for spark ignition engines. This
result might, at first, seem to contradict the pattern in spark ignition engines, for
which NO_x emissions are worst for an equivalence ratio of about 0.9. In diffusion
flames, fuel is diffusing towards the oxidant, and oxidant diffuses towards the
fuel. The equivalence ratio varies continuously, from high values at the fuel
droplet to values less than unity in the surrounding gases. The flame position can
be defined for mathematical purposes as where the equivalence ratio is unity.
However, the reaction zone will extend each side of the stoichiometric region to
wherever the mixture is within the flammability limits. This will have an averag-
ing effect on NO_x production. In addition, radiation from the reaction zone is
significant, and NO_x production is strongly temperature-dependent. A common
method to reduce NO_x emissions is to retard the injection timing, but this has
adverse effects on fuel consumption and smoke emissions. Retarding the injection
timing may be beneficial because this reduces the delay period and consequently
the uncontrolled combustion period.

The most serious emission from compression ignition engines is smoke, with
the characteristic grey or black of soot (carbon) particles. In this discussion,
smoke does not include the bluish smoke that signifies lubricating oil is being
burnt, or the white smoke that is characteristic of unburnt fuel. These types of

smoke occur only with malfunctioning engines, both compression and spark ignition.

Smoke from compression ignition engines originates from carbon particles formed by cracking (splitting) of large hydrocarbon molecules on the fuel-rich side of the reaction zone. The carbon particles can grow by agglomeration until they reach the fuel-lean zone, where they can be oxidised. The final rate of soot release depends on the difference between the rate of formation and the rate of oxidation. The maximum fuel injected (and consequently power output) is limited so that the exhaust smoke is just visible. Smoke output can be reduced by advancing the injection timing or by injecting a finer fuel spray, the latter being obtained by higher injection pressures and finer nozzles. Smoke from a compression ignition engine implies a poorly calibrated injector pump or faulty injectors.

3.9 Combustion modelling

3.9.1 Introduction

The combustion model is one of the key elements in any computer simulation of internal combustion engine cycles. In addition, all aspects of the engine operating cycle directly influence the combustion process. Heywood (1980) provides a very good introduction to the subject and he emphasises the inter-dependence and complication of the combustion and engine operation. The combustion occurs in a three-dimensional, time-dependent, turbulent flow, with a fuel containing a blend of hydrocarbons, and with poorly understood combustion chemistry. The combustion chamber varies in shape, and the heat transfer is difficult to predict.

There are three approaches to combustion modelling; in order of increasing complexity, they are:

(i)   Zero-dimensional models (or phenomenological models). These use an empirical 'heat release' model, in which time is the only independent variable.

(ii) Quasi-dimensional models. These use a separate submodel for turbulent combustion to derive a 'heat release' model.

(iii) Multi-dimensional models. These models solve numerically the equations for mass, momentum, energy and species conservation in one, two or three dimensions, in order to predict the flame propagation.

All models can be used for estimating engine efficiency, performance, and emissions. The zero-dimensional and quasi-dimensional models are readily incorporated into complete engine models, but there is no explicit link with combustion chamber geometry. Consequently, these models are useful for parametric studies associated with engine development. When combustion chamber geometry is important or subject to much change, multi-dimensional models have to be used. Since the computational demands are very high, multi-dimensional models are used for combustion chamber modelling rather than complete engine modelling.

The more complex models are still subject to much research and refinement, and rely on submodels for the turbulence effects and chemical kinetics. Review papers by Tabaczynski (1983) and by Greenhalgh (1983) illustrate the use of lasers in turbulence (laser doppler anemometry/velocimetry — LDA, LDV) and the use of lasers in chemical species measurements (spectrographic techniques), respectively. These techniques can be applied to operating engines fitted with quartz windows for optical access. All models require experimental validation with engines, and combustion films can be invaluable for checking combustion models.

3.9.2 Zero-dimensional models

This approach to combustion modelling is best explained by reference to a particular model, the one described by Heywood *et al.* (1979), for spark ignition engines. This combustion model makes use of three zones, two of which are burnt gas:

(i)   unburnt gas
(ii) burnt gas
(iii) burnt gas adjacent to the combustion chamber — a thermal boundary layer or quench layer.

This arrangement is shown in figure 3.15, in addition to the reaction zone or flame front separating the burnt and unburnt gases. The combustion does not occur instantaneously, and can be modelled by a Wiebe function:

$$x(\theta) = 1 - \exp\{-a\,[(\theta - \theta_0)/\Delta\theta_b]\,m + 1\}$$

where $x(\theta)$ is the mass fraction burnt at crank angle θ
                 θ_0 is the crank angle at the start of combustion
and $\Delta\theta_b$ is the duration of combustion.

a and m are constants that can be varied so that a computed p-V diagram can be matched to that of a particular engine. Typically

$$a = 5 \quad \text{and} \quad m = 2$$

The effect of varying these parameters on the rate of combustion is shown in figure 3.16.

The heat transfer is predicted using the correlation developed by Woschni (1968); although this was developed for compression ignition engines, it is

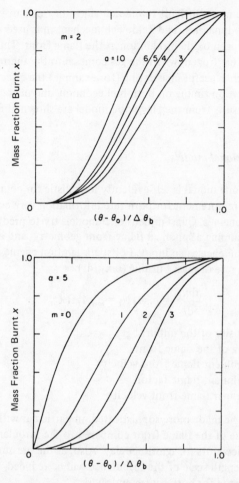

Figure 3.16   Wiebe functions (reprinted with permission from Heywood *et al.*
(1979), © Society of Automotive Engineers, Inc.)

widely used for spark ignition engines. The correlation has a familiar form, in
terms of Nusselt, Reynolds and Prandtl numbers:

$$Nu = a\,Re^b Pr^c$$

The constants a, b and c will depend on the engine geometry and speed but
typical values are

$$a = 0.035, b = 0.8, c = 0.333$$

As well as predicting the engine efficiency, this type of model is very useful
in predicting engine emissions. The concentrations of carbon-oxygen-hydrogen
species in the burnt gas are calculated using equilibrium thermodynamics. Nitric

oxide emissions are more difficult to predict, since they cannot be described by equilibrium thermodynamics. The Zeldovich mechanism is used as a basis for calculating the nitrogen oxide production at the flame front. This composition is then assumed to be 'frozen' when the gas comes into the thermal boundary layer. This approach is useful in assessing (for example) the effect of exhaust gas recirculation or ignition timing on both fuel economy and nitrogen oxide emissions. Some results from the Heywood model are shown in figure 3.17.

3.9.3 Quasi-dimensional models

The simple three zone model is self-evidently unrealistic for compression ignition engines; nor is the requirement for burn rate information very convenient — even for spark ignition engines. Quasi-dimensional models try to predict the burn rate information by assuming a spherical flame front geometry, and by using information about the turbulence as an input. For spark ignition engines this simple approach gives the rate of mass burning (dm_b/dt) as

$$\frac{dm_b}{dt} = \rho_u A_f U_t = \rho_u A_f \text{ff } U_1$$

where ρ_u = density of the unburnt gas
A_f = area of the flame front
U_t = turbulent flame front velocity
ff = turbulent flame factor
U_1 = laminar flame front velocity.

This approach can be made more sophisticated, in particular with regard to the turbulence. The size of the flame front compared to the turbulence scale changes, for example; this accounts for ignition delay. Also, as the pressure rises during compression, the length scale of the turbulence will be reduced. Finally, allowance needs to be made for both squish and swirl.

For compression ignition engines, the mixing of the air and the fuel jet are all important, and a turbulent jet-entrainment model is necessary. Since the time histories of different fuel elements will not be the same, a multi-zone combustion model is needed to trace the individual fuel elements. The prediction of NO_x emissions will provide a powerful check on any such model, since NO_x production will be very sensitive to the wide variations in both temperature and air/fuel ratio that occur in compression ignition engines.

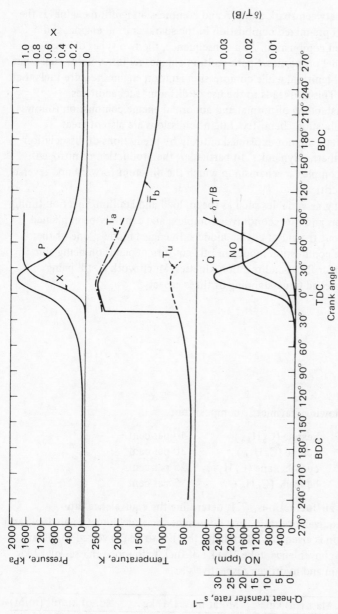

Figure 3.17   Profiles of variables predicted by the simulation throughout the four-stroke engine cycle for 5.7 litre displacement engine at base operating point. Plotted against crank angle are: mass fraction burnt x, unburnt mixture temperature T_u, mean burnt gas temperature $\overline{T}_b$, cylinder pressure p, temperature of burnt gas adiabatic core T_a, instantaneous heat transfer rate $\dot{Q}$ (normalised by the initial enthalpy of the fuel/air mixture within the cylinder), nitric oxide concentration NO, and thermal boundary layer thickness δ_T (normalised by the cylinder bore B) (reprinted with permission from Heywood *et al.* (1979), © Society of Automotive Engineers, Inc.)

3.10 Conclusions

A key difference between spark ignition and compression ignition engines is the difference between pre-mixed combustion in the spark ignition engine, and diffusion-controlled combustion. As a consequence, the two types of engine require different fuel properties. Spark ignition engines require volatile fuels that are resistant to self-ignition, while compression ignition engines require fuels that self-ignite readily. This also leads to the use of different fuel additives.

A proper understanding of normal and abnormal engine combustion follows from fuel and combustion chemistry. Engine emissions are also of great importance, but these are not explained entirely by predictions of dissociation from equilibrium thermodynamics. In particular, the production of nitrogen oxides involves a complex mechanism in which the different forward and reverse reaction rates are critical.

The effect of engine variables such as speed, load and ignition/injection timing on engine emissions and fuel economy is complex, and can only be explained qualitatively without the aid of engine models. In engine models, one of the critical parts is the combustion model and this in turn depends on being able to model the in-cylinder flows and reaction kinetics. Much work is still being devoted to an improved understanding of these aspects.

3.11 Examples

Example 3.1

A fuel has the following gravimetric composition;

hexane (C_6H_{14})	40 per cent
octane (C_8H_{18})	30 per cent
cyclohexane (C_6H_{12})	25 per cent
benzene (C_6H_6)	5 per cent

If the gravimetric air/fuel ratio is 17:1, determine the equivalence ratio.
To calculate the equivalence ratio, first determine the stoichiometric air/fuel ratio. As the composition is given gravimetrically, this has to be converted to molar composition. For convenience, take 100 kg of fuel, with molar masses of 12 kg and 1 kg for carbon and hydrogen respectively.

Substance	Mass (m) kg	Molar mass (M) kg	No. of kmols (m/M)
C_6H_{14}	40	(6.12 + 14.1)	0.465
C_8H_{18}	30	(8.12 + 18.1)	0.263
C_6H_{12}	25	(6.12 + 12.1)	0.298
C_6H_6	5	(6.12 + 6.1)	0.064

Assuming that the combustion products contain only CO_2 and H_2O, the stoichiometric reactions in terms of kmols are found to be

$$C_6H_{14} + x \left(O_2 + \frac{79}{21} N_2^* \right) \rightarrow 6CO_2 + 7H_2O + yN_2^*$$

Variable x is found from the number of kilomoles of oxygen associated with the products, $x = 9\frac{1}{2}$, and by conservation of N_2^*, $y = (79/21)x$:

$$C_6H_{14} + 9\frac{1}{2} \left(O_2 + \frac{79}{21} N_2^* \right) \rightarrow 6CO_2 + 7H_2O + 35.74N_2^*$$

1 kmol of fuel: $9\frac{1}{2} \left(1 + \frac{79}{21} \right)$ kmols of air = 1 : 45.23

By inspection:

$$C_8H_{18} + 12\frac{1}{2} \left(O_2 + \frac{79}{21} N_2^* \right) \rightarrow 8CO_2 + 9H_2O + 47.0N_2^*$$

1 : 59.52 kmols

$$C_6H_{12} + 9 \left(O_2 + \frac{79}{21} N_2^* \right) \rightarrow 6CO_2 + 6H_2O + 33.86N_2^*$$

1 : 42.86 kmols

$$C_6H_6 + 7\frac{1}{2} \left(O_2 + \frac{79}{21} N_2^* \right) \rightarrow 6CO_2 + 3H_2O + 28.21N_2^*$$

1 : 35.71 kmols

Fuel	kmols of fuel	Stoichiometric molar air/fuel ratio	kmols of air
C_6H_{14}	0.465	45.23:1	21.03
C_8H_{18}	0.263	59.52:1	15.65
C_6H_{12}	0.298	42.86:1	12.77
C_6H_6	0.064	35.71:1	2.29
Total	100 kg fuel		51.74

100 kg fuel : 51.74 × 29 kg air

1 : 15.00

Equivalence ratio, $\phi = \dfrac{\text{(air/fuel ratio) stoichiometric}}{\text{(air/fuel ratio) actual}}$

$$\phi = \frac{15}{17} = 0.882$$

Example 3.2

A fuel oil has a composition by weight of 0.865 carbon, 0.133 hydrogen and 0.002 incombustibles. Find the stoichiometric gravimetric air/fuel ratio.
When the fuel is burnt with excess air, the dry volumetric exhaust gas analysis is: CO_2 0.121, N_2^* 0.835, O_2 0.044. Determine the actual air/fuel ratio used and the wet volumetric exhaust gas analysis.
Consider 1 kg of fuel, and convert the gravimetric data to molar data. Molar mass of carbon = 12, hydrogen = 1.

$$\frac{0.865}{12} C \; + \; \frac{0.133}{1} H + Air \qquad\qquad \rightarrow Products$$

$$0.0721C \; + 0.133H \; + a \left(O_2 + \frac{79}{21} N_2^*\right) \; \rightarrow 0.0721CO_2 + 0.0665H_2O + 0.396N_2^*$$

For a stoichiometric reaction, $a = 0.0721 + \frac{1}{2}(0.0665) = 0.105$.

Thus 1 kg of fuel combines with $0.105 \left(1 + \frac{79}{21}\right)$ kmol of air

$$\text{or } 0.105 \left(1 + \frac{79}{21}\right) \times 29 \text{ kg of air}$$

and the stoichiometric gravimetric air/fuel ratio is 14.5 : 1.
Note that the molar mass of the fuel is not known.

With excess air the equation can be rewritten with variable x. The incombustible material is assumed to occupy negligible volume in the products. A dry gas analysis assumes that any water vapour present in the combustion products has been removed.

$$0.0721C + 0.133H + 0.105x\, O_2 + 0.396x\, N_2^* \rightarrow 0.0721CO_2 + 0.0665H_2O$$
$$+ 0.396x\, N_2^* + 0.105(x - 1)O_2$$

The variable x is unity for stoichiometric reactions; if $x > 1$ there will be $0.105(x - 1)$ kmols of O_2 not taking part in the reaction with 1 kg fuel.
The dry volumetric gas analysis for each constituent in the products is in proportion to the number of moles of each dry constituent:

$$CO_2: 0.121 = \frac{0.0721}{0.0721 + 0.396x + 0.105(x - 1)}$$

$$0.0721/0.121 \; = 0.0721 + (0.396 + 0.105)x - 0.105; \text{ hence } x = 1.255.$$

$$O_2 \; : 0.044 = \frac{0.105(x - 1)}{0.0721 + 0.396x + 0.105(x - 1)}$$

$0.105x - 0.044(0.396 + 0.105)x = 0.044(0.0721 - 0.105) + 0.105;$

$$\text{hence } x = 1.248.$$

$$N_2^* : 0.835 = \frac{0.396x}{0.0721 + 0.396x + 0.105(x - 1)}$$

$$\left(0.396 + 0.105 - \frac{0.396}{0.835}\right)x = 0.105 - 0.0721; \text{ hence } x = 1.230.$$

As might be expected, each value of x is different. The carbon balance is most satisfactory since it is the largest measured constituent of the exhaust gases. The nitrogen balance is the least accurate, since it is found by difference $(0.835 = 1 - 0.044 - 0.121)$. The redundancy in these equations can also be used to determine the gravimetric composition of the fuel.

Taking $x = 1.25$ the actual gravimetric air/fuel ratio is

$$14.5 \times 1.25 : 1 = 18.13 : 1$$

The combustion equation is actually

$$0.0721C + 0.0133H + 0.13130_2 + 0.495N_2^* \rightarrow 0.0721CO_2 + 0.0665H_2O$$
$$+ 0.495N_2^* + 0.0263O_2$$

The wet volumetric analysis for each constituent in the products is in proportion to the number of moles of each constituent, including the water vapour

$$CO_2: \frac{0.0721}{0.0721 + 0.0665 + 0.495 + 0.0263} = \frac{0.0721}{0.6599} = 0.1093$$

$$H_2O: \frac{0.0665}{0.6599} = 0.1008$$

$$N_2^*: \frac{0.495}{0.6599} = 0.7501$$

$$O_2: \frac{0.0263}{0.6599} = 0.0399$$

Check that the sum equals unity $(0.1093 + 0.1008 + 0.7501 + 0.0399 = 1.0001)$.

Example 3.3

Calculate the difference between the constant-pressure and constant-volume lower calorific values (LCV) for ethylene (C_2H_4) at 250°C and 1 bar. Calculate the constant-pressure calorific value at 2000 K. What is the constant-pressure higher calorific value (HCV) at 25°C?
The ethylene is gaseous at 25°C, and the necessary data are in the tables by Rogers and Mayhew (1980b):

C_2H_4 (vap.) $+ 3O_2 \rightarrow 2CO_2 + 2H_2O$ (vap.), $(\Delta H_0)_{25°C} = -1\ 323\ 170$ kJ/kmol

(i)   Using equation (3.4):

$$(-\Delta H_0)_{25°C} - (-\Delta U_0)_{25°C} = (n_{GR} - n_{GP})R_0 T$$

At 25°C the partial pressure of water vapour is 0.032 bar, so it is fairly accurate to assume that all the water vapour condenses; thus

n_{GR}, number of moles of gaseous reactants $= 1 + 3$

n_{PR}, number of moles of gaseous products $= 2 + 0$

$$(n_{GR} - n_{GP})R_0 T = (4 - 2) \times 8.3144 \times 298.15 = 4.958 \text{ kJ/kmol}$$

This difference in calorific values can be seen to be negligible. If the reaction occurred with excess air the result would be the same, as the atmospheric nitrogen and excess oxygen take no part in the reaction.

(ii)   To calculate $(-\Delta H_0)_{2000\,K}$, use can be made of equation (3.3); this amounts to the same as the following approach:

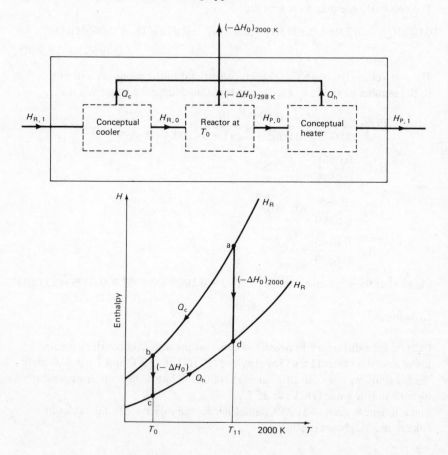

$$(-\Delta H_0)_{2000\,K} = Q_c + (-\Delta H_0)_{298\,K} + Q_h$$

$$a \rightarrow d \equiv a \rightarrow b \rightarrow c \rightarrow d$$

$$Q_c = (H_{2000\,K} - H_{298\,K})_{C_2H_4} + 3(H_{2000\,K} - H_{298\,K})_{O_2}$$

$$= (159\,390 - 0) + 3(59\,199 - 0)$$

$$= 336\,987$$

$$Q_R = 2(H_{2000\,K} - H_{298\,K})_{CO_2} + 2(H_{2000\,K} - H_{298\,K})_{H_2O}$$

$$= 2(91\,450 - 0) + 2(72\,689 - 0) = 328\,278$$

$$(HCV)_{2000\,K} = (-\Delta H_0)_{2000\,K} = 336\,987 + 1\,323\,170 - 328\,278$$

$$= \underline{1\,331\,879 \text{ kJ/kmol}}$$

Again this is not a significant difference. As before, if the reaction occurred with excess air the result would be the same, as the excess oxygen and atmospheric nitrogen take no part in the reaction, and have the same initial and final state.

(iii)   As the enthalpy of reaction is for water in the vapour state, this corresponds to the lower calorific value.

Using equation (3.5)

$$HCV - LCV = y\, h_{fg}$$

where y = mass of water vapour/kmol of fuel
or

$$HCV - LCV = z\, H_{fg}$$

where z = no. of kmols of water vapour/kmol fuel.

$$H_{fg} = 43\,990 \text{ kJ/kmol of } H_2O \text{ at } 25°C$$

Thus

$$HCV = LCV + 2H_{fg}$$

$$= 1\,323\,170 + 2 \times 43\,990 = 1\,411\,150 \text{ kJ/kmol fuel}$$

Example 3.4

In a closed combustion vessel propane (C_3H_8) and air with an equivalence ratio of 1.11 initially at 25°C burn to produce products consisting solely of carbon dioxide (CO_2), carbon monoxide (CO), water (H_2O) and atmospheric nitrogen. If the heat rejected from the vessel is 770 MJ per kmol of fuel, show that the final temperature is about 1530°C.
If the initial pressure is 1 bar, estimate the final pressure.
Lower calorific value of propane $(-\Delta H_0)_a$, is 2 039 840 kJ/kmol of fuel at 25°C.

As no oxygen is present in the combustion products, this implies that dissociation should be neglected. The stoichiometric reaction is

$$C_3H_8 + 5\left(O_2 + \frac{79}{21}N_2^*\right) \rightarrow 3CO_2 + 4H_2O + 5 \times \frac{79}{21}N_2^*$$

In this problem the equivalence ratio is 1.11, thus

$$C_3H_8 + 4.5\left(O_2 + \frac{79}{21}N_2^*\right) \rightarrow 2CO_2 + CO + 4H_2O + 16.93N_2^*$$

Let the enthalpy of reaction be $(\Delta H_0)_b$ for this reaction.
If $(\Delta H_0)_c$ is the enthalpy of reaction for

$$CO + \tfrac{1}{2}O_2 \rightarrow CO_2$$

then $(\Delta H_0)_b = (\Delta H_0)_a - (\Delta H_0)_c$

Using data from Rogers and Mayhew (1980b)

$$(\Delta H_0)_b = -2\,039\,840 - (-282\,990) = -1\,756\,850 \text{ kJ/kmol of fuel at } 25°C$$

Equation (3.4) converts constant-pressure to constant-volume calorific values:

$$(CV)_V = (CV)_p - (n_{GR} - n_{GP})R_0T$$
$$= 1\,756\,850 - \{(5.5 - 7) \times 8.3144 \times 298.15\}$$
$$= 1\,760\,568 \text{ kJ/kmol fuel at } 25°C$$

Assume the final temperature is 1800 K and determine the heat flow. Applying the 1st Law of Thermodynamics

$$(CV)_V = 2(U_{1800} - U_{298})_{CO_2} + 1(U_{1800} - U_{298})_{CO} + 4(U_{1800} - U_{298})_{H_2O}$$
$$+ 16.93(U_{1800} - U_{298})_{N_2^*} - Q$$
$$1\,760\,568 = 2(64\,476 + 2479) + 1(34\,556 + 2479) + 4(47\,643 + 2479)$$
$$+ 16.93(34\,016 + 2479) - Q$$

$$Q = -771\,275 \text{ kJ/kmol fuel}$$

The negative sign indicates a heat flow from the vessel, and the numerical value is sufficiently close to 770 MJ.

To estimate the final pressure, apply the equation of state; use a basis of 1 kmol of fuel.

$$pV = nR_0T$$

Find V for the reactants

$$V = \frac{n_R R_0 T}{p} = \frac{(1 + 4.5 + 1693)R_0\,298}{10^5}$$

for the products

$$p = \frac{n_R R_0 T}{V}$$

$$= \frac{(2 + 1 + 4 + 16.93) R_0}{(1 + 4.5 + 16.93) R_0} \frac{1800}{298} \times 10^5$$

$$= \underline{6.53 \text{ bar}}$$

Example 3.5

Compute the partial pressures of a stoichiometric equilibrium mixture of CO, O_2, CO_2 at 3000 and 3500 K when the pressure is 1 bar. Show that for a stoichiometric mixture of CO and O_2 initially at 25°C and 1 bar, the adiabatic constant-pressure combustion temperature is about 3050 K.

Introduce a variable x into the combustion equation to account for dissociation

$$CO + \tfrac{1}{2}O_2 \rightarrow (1 - x)CO_2 + xCO + \frac{x}{2}O_2$$

The equilibrium constant K is defined by equation (3.7):

$$K = \frac{p'_{CO_2}}{(p'_{CO})(p'_{O_2})^{\frac{1}{2}}} \text{ where } p'_{CO_2} \text{ is the partial pressure of } CO_2 \text{ etc.}$$

The partial pressure of a species is proportional to the number of moles of that species. Thus

$$p'_{CO_2} = \frac{1 - x}{(1 - x) + x + \dfrac{x}{2}} p, \quad p'_{CO} = \frac{x}{1 + \dfrac{x}{2}} p, \quad p'_{O_2} = \frac{x/2}{1 + \dfrac{x}{2}} p$$

where p is the total system pressure.
Thus

$$K = \frac{1 - x}{x(x/2)^{\frac{1}{2}}} \left(\frac{1 + (x/2)}{p}\right)^{\frac{1}{2}}$$

Again, data are obtained from Rogers and Mayhew (1980b).
At 3000 K, $\log_{10} K = 0.485$ with pressure in atmospheres:

$$1 \text{ atmosphere} = 1.01325 \text{ bar}$$

Thus

$$10^{0.4825} = \frac{1 - x}{x \sqrt{\left(\dfrac{x}{2}\right)}} \sqrt{\left(\frac{1 + x/2}{1.01325}\right)}$$

$$10^{0.4825}(1.01325)^{-\frac{1}{2}} x^{\frac{3}{2}} = \sqrt{(2)}\,(1-x)\,\sqrt{(1+(x/2))}$$

Square all terms

$$10^{0.965}(1.01325)^{-1} x^3 = (1-x)^2\,(2+x)$$

$$8.211x^3 + 3x - 2 = 0$$

This cubic equation has to be solved iteratively. The Newton–Raphson method provides fast convergence; an alternative simpler method, which may not converge, is the following method

$$x_{n+1} = 3\sqrt{\left/\left(\frac{2-3x_n}{8.211}\right)\right.}\quad \text{where } n \text{ is the iteration number}$$

Using a calculator without memory this readily gives

$$x = 0.436$$

At 3500 K use linear interpolation to give $\log_{10} K = -0.187$, leading to

$$x = 0.748$$

This implies greater dissociation at the higher temperature, a result predicted by Le Châtelier's Principle for an exothermic reaction.

The partial pressures are found by back-substitution, giving the results tabulated below:

T	p	x	p'_{CO_2}	p'_{O_2}	p'_{CO}	
3000 K	1	0.436	0.463	0.179	0.358	all pressures in bar
3500 K	1	0.748	0.183	0.272	0.544	

To find the adiabatic combustion temperature, assume a temperature and then find the heat transfer necessary for the temperature to be attained. For the reaction $CO + \frac{1}{2}O_2 \rightarrow CO_2$, $(\Delta H_0)_{298} = -282\,990$ kJ/kmol.
Allowing for dissociation, the enthalpy of reaction is $(1-x)(\Delta H_0)_{298}$.

It is convenient to use the following hypothetical scheme for the combustion calculation

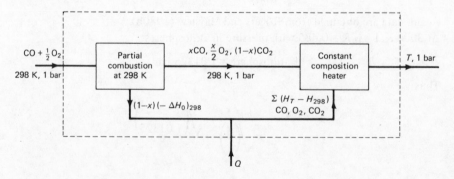

$$(1-x)(-\Delta H_0)_{298} + Q = \sum_{CO,O_2,CO_2} (H_T - H_{298})$$

If temperature T is guessed correctly then $Q = 0$.
Alternatively, make two estimates and interpolate.
1st guess: $T = 3000$ K, $x = 0.436$.

$$Q = \left\{(1-x)(H_{3000})_{CO_2} + x(H_{3000})_{CO} + \frac{x}{2}(H_{3000})_{O_2}\right\} -(1-x)(-\Delta H_0)_{298};$$

$H_{298} = 0$

$= \{0.564 \times 152\,860 + 0.436 \times 93\,542 + 0.218 \times 98\,098\} - 0.564 \times 282\,990$

$= -11\,223$ kJ/kmol CO, that is, T is too low.

2nd guess: $T = 3500$ K, $x = 0.748$.

$Q = \{0.252 \times 184\,130 + 0.748 \times 112\,230 + 0.374 \times 118\,315\} - 0.252 \times 282\,290$

$= 103\,462$ kJ/kmol CO, that is, T is much too high.

To obtain a better estimate of T, interpolate between the 1st and 2nd guesses.

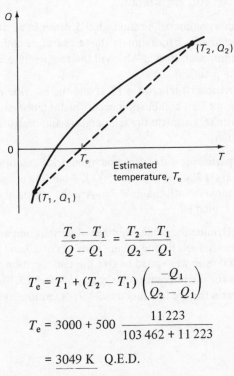

$$\frac{T_e - T_1}{Q - Q_1} = \frac{T_2 - T_1}{Q_2 - Q_1}$$

$$T_e = T_1 + (T_2 - T_1)\left(\frac{-Q_1}{Q_2 - Q_1}\right)$$

$$T_e = 3000 + 500 \frac{11\,223}{103\,462 + 11\,223}$$

$$= 3049 \text{ K} \quad \text{Q.E.D.}$$

If a more accurate result is needed, recalculate Q when $T = 3050$ K. Assuming no dissociation, $T \approx 5000$ K.

3.12 Problems

3.1 If a fuel mixture can be represented by the general formula C_xH_{2x}, show that the stoichiometric gravimetric air/fuel ratio is 14.8:1.

3.2 A fuel has the following molecular gravimetric composition

pentane (C_5H_{12})	10 per cent
heptane (C_7H_{16})	30 per cent
octane (C_8H_{18})	35 per cent
dodecane ($C_{12}H_{26}$)	15 per cent
benzene (C_6H_6)	10 per cent

Calculate the atomic gravimetric composition of the fuel and the gravimetric air/fuel ratio for an equivalence ratio of 1.1.

3.3 The dry exhaust gas analysis from an engine burning a hydrocarbon diesel fuel is as follows: CO_2 0.121, O_2 0.037, N_2^* 0.842. Determine the gravimetric composition of the fuel, the equivalence ratio of the fuel/air mixture, and the stoichiometric air/fuel ratio.

3.4 An engine with a compression ratio of 8.9:1 draws in air at a temperature of 20°C and pressure of 1 bar. Estimate the temperature and pressure at the end of the compression stroke. Why will the temperature and pressure be less than this in practice?

    The gravimetric air/fuel ratio is 12:1, and the calorific value of the fuel is 44 MJ/kg. Assume that combustion occurs instantaneously at the end of the compression stroke. Estimate the temperature and pressure immediately after combustion.

3.5 Compute the partial pressures of a stoichiometric equilibrium mixture of CO, O_2, CO_2 at (i) 3000 K and (ii) 3500 K when the pressure is 10 bar. Compare the answers with example 3.5. Are the results in accordance with Le Châtelier's Principle?

3.6 In a test to determine the cetane number of a fuel, comparison was made with two reference fuels having cetane numbers of 50 and 55. In the test the compression ratio was varied to give the same ignition delay. The compression ratios for the reference fuels were 25.4 and 23.1 respectively. If the compression ratio for the test fuel was 24.9, determine its cetane number.

3.7 Contrast combustion in compression ignition engines and spark ignition engines. What are the main differences in fuel requirements?

3.8 What is the difference between 'knock' in compression ignition and spark ignition engines? How can 'knock' be eliminated in each case?

4 Spark Ignition Engines

4.1 Introduction

This chapter considers how the combustion process is initiated and constrained in spark ignition engines. The air/fuel mixture has to be close to stoichiometric (chemically correct) for satisfactory spark ignition and flame propagation. The equivalence ratio or mixture strength of the air/fuel mixture also affects pollutant emissions, as discussed in chapter 3, and influences the susceptibility to spontaneous self-ignition (that is, knock). A lean air/fuel mixture (equivalence ratio less than unity) will burn more slowly and will have a lower maximum temperature than a less lean mixture. Slower combustion will lead to lower peak pressures, and both this and the lower peak temperature will reduce the tendency for knock to occur. The air/fuel mixture also affects the engine efficiency and power output. At constant engine speed with fixed throttle, it can be seen how the brake specific fuel consumption (inverse of efficiency) and power output vary. This is shown in figure 4.1 for a typical spark ignition engine at full or wide open throttle (WOT). As this is a constant-speed test, power output is proportional to torque output, and this is most conveniently expressed as bmep since bmep is independent of engine size. Figure 4.2 is an alternative way of expressing the same data (because of their shape, the plots are often referred to as 'fish-hook' curves); additional part throttle data have also been included. At full throttle, the maximum for power output is fairly flat, so beyond a certain point a richer mixture significantly reduces efficiency without substantially increasing power output. The weakest mixture for maximum power (wmmp) will be an arbitrary point just on the lean side of the mixture for maximum power. The minimum specific fuel consumption also occurs over a fairly wide range of mixture strengths. A simplified explanation for this is as follows. With dissociation occurring, maximum power will be with a rich mixture when as much as possible of the oxygen is consumed; this implies unburnt fuel and reduced efficiency. Conversely, for maximum economy as much of the fuel should be burnt as possible, implying a weak mixture with excess oxygen present. In addition, it was shown in chapter 2 that the weaker the air/fuel mixture the higher the ideal

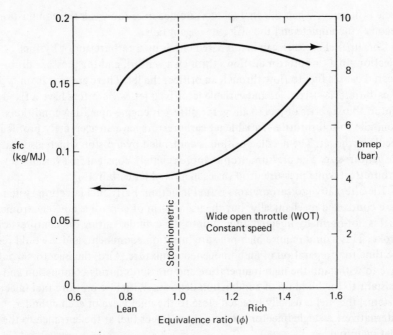

Figure 4.1 Response of specific fuel consumption and power output to changes in air/fuel ratio

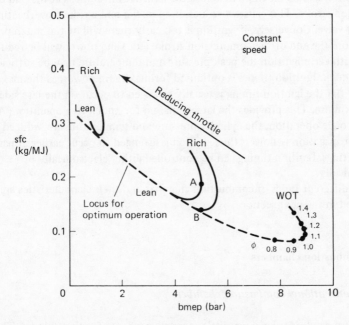

Figure 4.2 Specific fuel consumption plotted against power output for varying air/fuel ratios at different throttle settings

cycle efficiency. When the air/fuel mixture becomes too weak the combustion becomes incomplete and the efficiency again falls.

The air/fuel mixture can be prepared either by a carburettor or by fuel injection. In a carburettor air flows through a venturi, and the pressure drop created causes fuel to flow through an orifice, the jet. There are two main types of carburettor — fixed jet and variable jet. Fixed jet carburettors have a fixed venturi, but a series of jets to allow for different engine-operating conditions from idle to full throttle. Variable jet carburettors have an accurately profiled needle in the jet. The needle position is controlled by a piston which also varies the venturi size. The pressure drop is approximately constant in a variable jet carburettor, while pressure drop varies in a fixed jet carburettor.

The alternative to carburettors is fuel injection. Early fuel injection systems were controlled mechanically, but the usual form of control is now electronic. Fuel is not normally injected directly into the cylinder during the compression stroke. This would require high-pressure injection equipment, and it would reduce the time for preparation of an homogeneous mixture. Also, the injectors would have to withstand the high temperature and pressures during combustion and be resistant to the build-up of combustion deposits. With low-pressure fuel injection systems, the fuel is usually injected close to the inlet valve of each cylinder. Alternatively a single injector can be used to inject fuel at the entrance to the inlet manifold.

The ignition timing also has to be controlled accurately. If ignition is too late, combustion will be incomplete before the exhaust valve opens at the end of the expansion stroke. This causes a reduction in power and a risk of overheating the exhaust valve. Conversely, if ignition is too early there will be too much pressure rise before the end of the compression stroke (tdc) and power will be reduced. Also, with early ignition the peak pressure and temperature may be sufficient to cause knock. Ignition timing is optimised for maximum power; as the maximum is fairly flat the ignition timing is usually arranged to occur on the late side of the maximum. This provides the largest margin for knock-free operation. At part throttle operation, the cylinder pressure and temperature are reduced and flame propagation is slower; thus ignition is arranged to occur earlier at part load settings. Ignition timing can be controlled either electronically or mechanically.

The different types of combustion chamber and their characteristics are discussed in the next section.

4.2 Combustion chambers

4.2.1 Conventional combustion chambers

Initially the cylinder head was little more than a cover for the cylinder. The simplest configuration was the side valve engine, figure 4.3, with the inlet and

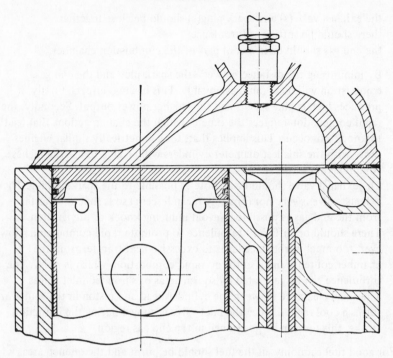

Figure 4.3 Ricardo turbulent head for side valve engines (reproduced with permission from Ricardo and Hempson (1968))

exhaust valves together at one side of the cylinder. The most successful combustion chamber for the side valve engine was the Ricardo turbulent head, as shown in figure 4.3. This design was the result of extensive experimental studies aimed at improving combustion. The maximum compression ratio that was reasonable with this geometry was limited to about 6:1, but this was not a restriction since the octane rating of fuels in the 1920s and 1930s was only about 60–70.

In the Ricardo turbulent head design the clearance between part of the cylinder head and piston at the end of the compression stroke is very small. This forms an area of 'squish', from which gas is ejected into the main volume. The turbulence that this jet generates ensures rapid combustion. If too large a squish area is used the combustion becomes too rapid and noisy. This design also reduces the susceptibility to knock, since the gas furthest from the sparking plug is in the squish area. The end gas in the squish area is less prone to knock since it will be cooler because of the close proximity of the cylinder head and piston. Excessive turbulence also causes excessive heat transfer to the combustion chamber walls, and should be avoided for this reason also.

The main considerations in combustion chamber design are:

(i)   the distance travelled by the flame front should be minimised

(ii)   the exhaust valve(s) and spark plug(s) should be close together
(iii) there should be sufficient turbulence
(iv)   the end gas should be in a cool part of the combustion chamber.

 (i)   By minimising the distance between the spark plug and the end gas,
       combustion will be as rapid as possible. This has two effects. Firstly, it
       produces high engine speeds and thus higher power output. Secondly, the
       rapid combustion reduces the time in which the chain reactions that lead
       to knock can occur. This implies that, for geometrically similar engines,
       those with the smallest diameter cylinders will be able to use the highest
       compression ratios.
(ii)   The exhaust valve should be as close as possible to the sparking plug. The
       exhaust valve is very hot (possibly incandescent) so it should be as far
       from the end gas as possible to avoid inducing knock or pre-ignition.
(iii) There should be sufficient turbulence to promote rapid combustion. How-
       ever, too much turbulence leads to excessive heat transfer from the
       chamber contents and also to too rapid combustion, which is noisy. The
       turbulence can be generated by squish areas or shrouded inlet valves.
(iv)   The small clearance between the cylinder head and piston in the squish area
       forms a cool region. Since the inlet valve is cooled during the induction
       stroke, this too can be positioned in the end gas region.

For good fuel economy all the fuel should be burnt and the quench areas where
the flame is extinguished should be minimised. The combustion chamber should
have a low surface-to-volume ratio to minimise heat transfer. The optimum swept
volume consistent with satisfactory operating speeds is about 500 cm^3 per cylinder. For high-performance engines, smaller cylinders will enable more rapid
combustion, so permitting higher operating speeds and consequently greater
power output. For a given geometry, reducing the swept volume per cylinder
from 500 cm^3 to 200 cm^3 might increase the maximum engine speed from
about 6000 rpm to 8000 rpm.

The ratio of cylinder diameter to piston stroke is also very important. When
the stroke is larger than the diameter, the engine is said to be 'under-square' In
Britain the car taxation system originally favoured under-square engines and this
hindered the development of higher-performance over-square engines. In over-
square engines the cylinder diameter is larger than the piston stroke, and this
permits larger valves for a given swept volume. This improves the induction and
exhaust processes, particularly at high engine speeds. In addition the short stroke
reduces the maximum piston speed at a given engine speed, so permitting higher
engine speeds. The disadvantage with over-square engines is that the combustion
chamber has a poor surface-to-volume ratio, so leading to increased heat transfer.
More recently there has been a return to under-square engines, as these have
combustion chambers with a better surface-to-volume ratio, and so lead to better
fuel economy.

Currently most engines have a compression ratio of about 9:1, for which a side valve geometry would be unsuitable. Overhead valve (ohv) engines have a better combustion chamber for these higher compression ratios. If the camshaft is carried in the cylinder block the valves are operated by push rods and rocker arms. A more direct alternative is to mount the camshaft in the cylinder head (ohc − overhead camshaft). The camshaft can be positioned directly over the valves, or to one side with valves operated by rocker arms. These alternatives are discussed more fully in chapter 6.

Figure 4.4a–d shows four fairly typical combustion chamber configurations; where only one valve is shown, the other is directly behind. Very often it will be production and economic considerations rather than thermodynamic considerations that determine the type of combustion chamber used. If combustion chambers have to be machined it will be cheapest to have a flat cylinder head and machined pistons. If the finish as cast is adequate, then the combustion chamber can be placed in the cylinder head economically.

Figure 4.4a shows a wedge combustion chamber; this is a simple chamber that produces good results. The valve drive train is easy to install, but the inlet and exhaust manifold have to be on the same side of the cylinder head. The hemispherical head, figure 4.4b has been popular for a long time in high-performance engines since it permits larger valves to be used than those with a flat cylinder head. The arrangement is inevitably expensive, with perhaps twin overhead camshafts. With the inlet and exhaust valves at opposite sides of the cylinder, it allows crossflow from inlet to exhaust. Crossflow occurs at the end of the exhaust stroke and the beginning of the induction stroke when both valves are open; it is beneficial since it reduces the exhaust gas residuals. More recently 'pent-roof' heads with four valves per cylinder have become popular; these have a shape similar to that of a house roof. The use of four valves gives an even greater valve area than does the use of two valves in a hemispherical head. A much cheaper alternative, which also has good performance, is the bowl in piston (Heron head) combustion chamber, figure 4.4c. This arrangement was used by Jaguar for their V12 engine and during development it was only marginally inferior to a hemispherical head engine with twin overhead camshafts (Mundy (1972)). The bath-tub combustion chamber, figure 4.4d, has a very compact combustion chamber that might be expected to give economical performance; it can also be used in a crossflow engine. All these combustion chambers have:

(i)   short maximum flame travel
(ii) the spark plug close to the exhaust valve
(iii) a squish area to generate turbulence
(iv) well-cooled end gas.

Figure 4.4e illustrates one of the new approaches to combustion chamber design. The provision of two sparking plugs and the geometry of the combustion chamber lead to a large flame front area. This leads to rapid combustion without the high heat transfer also associated with high turbulence. In any combustion chamber the highest combustion temperature will be from the mixture burnt last.

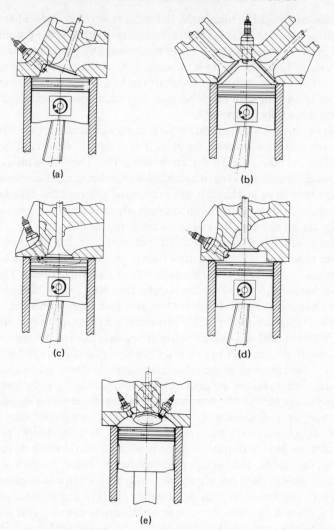

Figure 4.4   Combustion chambers for spark ignition engines. (a) Wedge chamber;
(b) hemispherical head; (c) bowl in piston chamber; (d) bath-tub
head; (e) Nissan NAPS-Z (or ZAPS) chamber

The reason is that the first burnt mixture compresses the unburnt gas, in addition
to any heat transfer from the burnt gas. Since the NAPS-Z combustion chamber
also utilises swirl, the last burnt gas will be in the centre of the combustion
chamber and consequently heat transfer from the combustion chamber will be
reduced.

The fuel economy of the spark ignition engine is particularly poor at part load;
this is shown in figure 4.2. Although operating an engine on a very lean mixture

can cause a reduction in efficiency, this reduction is less than if the power was controlled by throttling, with its ensuing losses. Too often, engine manufacturers are concerned with performance (maximum power and fuel economy) at or close to full throttle, although in automotive applications it is unusual to use maximum power except transiently.

In chapter 3 it was stated that the maximum compression ratio for an engine is usually dictated by the incipience of knock. If the problem of knock could be avoided, either by special fuels or special combustion chambers, there would still be an upper useful limit for compression ratio. As compression ratio is raised, there is a reduction in the rate at which the ideal cycle efficiency improves, see figure 4.5. Since the mechanical efficiency will be reduced by raising the compression ratio (owing to higher pressure loadings), the overall efficiency will be a maximum for some finite compression ratio, see figure 4.5.

Some of the extensive work by Caris and Nelson (1958) is summarised in figure 4.6. This work shows an optimum compression ratio of 16:1 for maximum economy, and 17:1 for maximum power. The reduction in efficiency is also due to poor combustion chamber shape — at high compression ratios there will be a poor surface-to-volume ratio. Figures for the optimum compression ratio vary since researchers have used different engines. For any given engine the optimum compression ratio will also be slightly dependent on speed, since mechanical efficiency depends on speed.

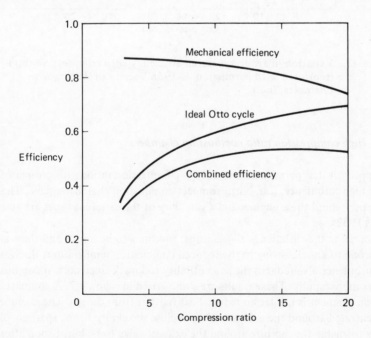

Figure 4.5 Variation in efficiency with compression ratio

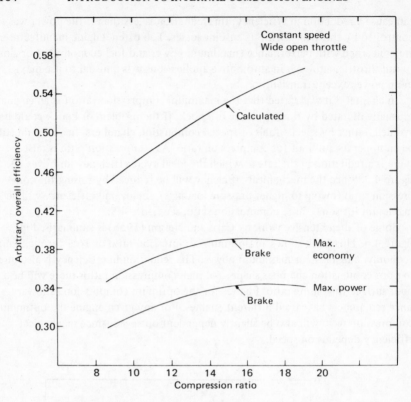

Figure 4.6   Variation in arbitrary overall efficiency with compression ratio
(reprinted with permission, © 1958 Society of Automotive
Engineers, Inc.)

4.2.2 High compression ratio combustion chambers

An approach that permits the use of high compression ratios with ordinary fuels
is the high turbulence, lean-burn, compact combustion chamber engine. The
concepts behind these engines and a summary of the different types are given by
Ford (1982).

Increasing the turbulence allows leaner mixtures to be burnt, and these are
less prone to knock, owing to the reduced combustion temperatures. Increasing
the turbulence also reduces the susceptibility to knock since normal combustion
occurs more rapidly. These results are summarised in figure 4.7. A compact combustion chamber is needed to reduce heat transfer from the gas. The chamber is
concentrated around the exhaust valve, in close proximity to the sparking plug;
this is to enable the mixture around the exhaust valve to be burnt soon after
ignition, otherwise the hot exhaust valve would make the combustion chamber

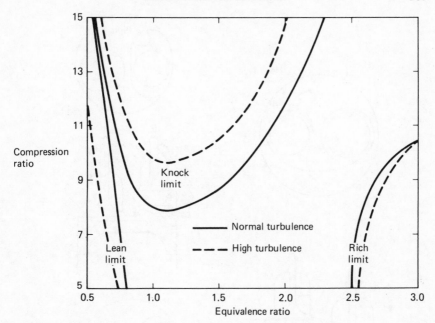

Figure 4.7 Effect of turbulence on increasing the operating envelope of spark ignition engines (adapted from Ford (1982))

prone to knock. The first design of this type was the May Fireball (May (1979)), with a flat piston and the combustion chamber in the cylinder head, figure 4.8a. Subsequently another design has been developed with the combustion chamber in the piston, and a flat cylinder head. In an engine with the compression ratio raised from 9.7:1 to 14.6:1, the gain in efficiency was up to 15 per cent at full throttle, with larger gains at part throttle.

The Ricardo High Ratio Compact Chamber (HRCC) design is very similar to the May Fireball, but has a straight passage from the inlet valve to the combustion chamber. The Ford Split Chamber is a more radical design, figure 4.8b. This design has a compact combustion chamber around the exhaust valve and a matching projection on the piston. As the piston approaches the end of the compression stroke the projection on the piston enters the combustion chamber. The combustion chamber is connected to the volume around the inlet valve only by a small passage, and as the piston completes its stroke a lot of turbulence is generated by the passage. Initial tests showed an improvement of 13–15 per cent in specific fuel consumption. There are also disadvantages associated with these combustion chambers. Emissions of carbon monoxide should be reduced, but hydrocarbon emissions (unburnt fuel) will be increased because of the large squish areas and poor surface-to-volume ratios. Hydrocarbon emissions can be removed by oxidation in a thermal reactor, with secondary air injection in the exhaust system. The exhaust temperatures will be low in this type of lean mixture engine, and the

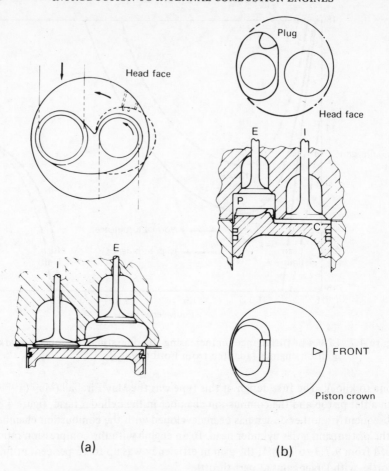

(a)   (b)

Figure 4.8   Compact high-turbulence combustion chambers for high compression
ratio engines burning lean mixtures. (a) May Fireball; (b) Ford Split
Chamber (from Ford (1982))

exhaust passages may have to be insulated to maintain sufficient temperature in
the thermal reactor. Emissions of nitrogen oxides (NO_x) will be greater for a
given air/fuel ratio, owing to the higher cylinder temperatures, but as the air/fuel
mixture will be leaner overall there should be a reduction in NO_x.

More careful control is needed on mixture strength and inter-cylinder distribution, in order to stay between the lean limit for misfiring and the limit for knock.
More accurate control is also needed on ignition timing. During manufacture
greater care is necessary, since tolerances that are acceptable for compression
ratios of 9:1 would be unacceptable at 14:1; in particular combustion chambers
need more accurate manufacture. Combustion deposits also have a more signifi-

cant effect with high compression ratios since they occupy a greater proportion
of the clearance volume.

4.3 Ignition systems

Most engines have a single sparking plug per cylinder, a notable exception being
in aircraft where the complete ignition system is duplicated to improve reliability.
The spark is usually provided by a battery and coil, though for some applications
a magneto is better. The timing of the spark is usually controlled mechanically.

For satisfactory performance, the central electrode of the sparking plug
should operate in the temperature range 350–700°C; if the electrode is too hot,
pre-ignition will occur. On the other hand, if the temperature is too low carbon
deposits will build up on the central insulator, so causing electrical breakdown.
The heat flows from the central electrode through the ceramic insulator; the
shape of this determines the operating temperature of the central electrode. A
cool-running engine requires a 'hot' or 'soft' sparking plug with a long heat flow
path in the central electrode, figure 4.9a. A hot-running engine, such as a high-
performance engine or a high compression ratio engine, requires a 'cool' or 'hard'
sparking plug. The much shorter heat flow path for a 'cool' sparking plug is shown
in figure 4.9b. The spark plug requires a voltage of 5–15 kV to spark; the larger
the electrode gap and the higher the cylinder pressure the greater the required
voltage.

Both a conventional coil ignition system and a magneto ignition system are
shown in figure 4.10. The coil is in effect a transformer with a primary or LT
(Low Tension) winding of about 200 turns, and a secondary or HT (High Tension) winding of about 20 000 turns of fine wire, all wrapped round an iron core.
The voltage V induced in the HT winding is

$$V = M \frac{\mathrm{d}I}{\mathrm{d}t} \qquad (4.1)$$

where I is the current flowing in the LT winding
        M is the mutual inductance $= k \sqrt{(L_1 L_2)}$
        L_1, L_2 are the inductances of the LT and HT windings, respectively
        (proportional to the number of turns squared)
and k is a coupling coefficient (less than unity)
or $V = k$ (turns ratio of windings) (Low Tension Voltage).

When the contact breaker closes to complete the circuit a voltage will be
induced in the HT windings, but it will be small since $\mathrm{d}I/\mathrm{d}t$ is limited by the
inductance and resistance of the LT winding. When the contact breaker opens
$\mathrm{d}I/\mathrm{d}t$ is much greater and sufficient voltage is generated in the HT windings to

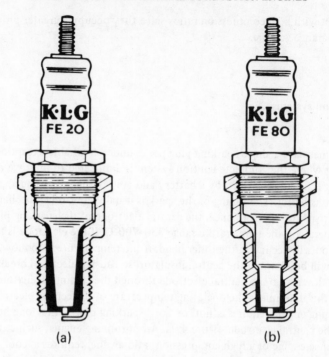

Figure 4.9   Sparking plugs. (a) Hot running; (b) cool running (from Campbell
            (1978))

jump the gaps between electrodes. A higher voltage (200–300 V) is generated in
the LT windings and this energy is stored in the capacitor. Without the capacitor
there would be severe arcing at the contact breaker. Once the spark has ended,
the capacitor discharges. The HT output is shown in figure 4.11. Initially 9 kV is
needed to ionise the gas sufficiently before the spark jumps with a voltage drop
of 2 kV. As engine speeds increase, the dwell period becomes shorter and the
spark energy will be reduced. Such a system can produce up to 400 sparks per
second; beyond this the spark energy becomes too low and this leads to misfiring.
For higher spark rates, twin coil/contact breaker systems or alternatively electronic
systems can be used. The current through the contact breaker can be reduced by
using it to switch the base of a transistor that controls the current to the LT
winding; this prolongs the life of the contact breaker. The contact breaker can
be replaced by making use of opto-electronic, inductive or magnetic switching.
However, all these systems use the coil in the same way as the contact breaker,
but are less prone to wear and maladjustment. Without the mechanical limitations of a contact breaker, a coil ignition system can produce up to 800 sparks
per second.

In a magneto there is no need to use a battery since a current is induced in the
LT winding by the changing magnetic field. Again a voltage is induced in the HT

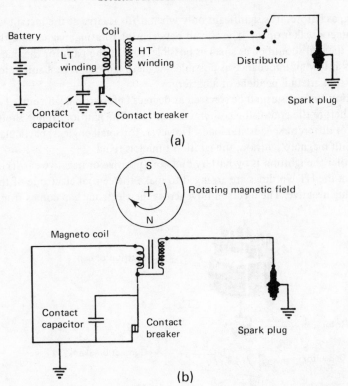

(a)

(b)

Figure 4.10 Mechanically operated ignition systems. (a) Conventional coil ignition system; (b) magneto ignition system (adapted from Campbell (1978))

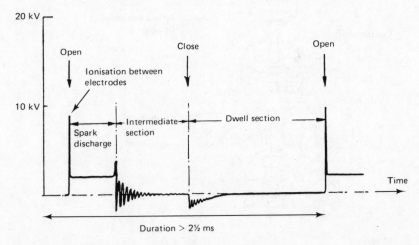

Figure 4.11 HT output from ignition coil (adapted from Campbell (1978))

winding; as before, it is significant only when dI/dt is large at the instant when the contact breaker opens. The air gap between the rotating magnet and the iron core of the coil should be as small as possible, so that the path for the magnetic flux has as low an impedance as possible. Magneto ignition is best suited to engines that are independent of a battery.

Ignition timing is usually expressed as degrees before top dead centre (°btdc), that is, before the end of the compression stroke. The ignition timing should be varied for different speeds and loads. However, for small engines, particularly those with magneto ignition, the ignition timing is fixed.

Whether the ignition is by battery and coil (positive or negative earth) or magneto, the HT windings are arranged to make the central electrode of the spark plug negative. The electron flow across the electrode gap comes from the

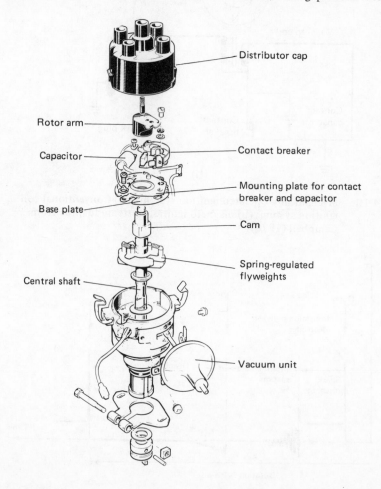

Figure 4.12   Automotive distributor (courtesy of Lucas Electrical Ltd)

negative electrode (the cathode), and the electrons flow more readily from a hot electrode. Since the central electrode is not in direct contact with the cylinder head, this is the hotter electrode. By arranging for the hotter electrode to be the cathode the breakdown voltage is reduced.

In chapter 3, section 3.5, it was explained how turbulent flame propagation occupies an approximately constant fraction of the engine cycle, since at higher speeds the increased turbulence gives a nearly corresponding increase in flame propagation rate. However, the initial period of flame growth occupies an approximately constant time ($\frac{1}{2}$ ms) and this corresponds to increased crank angles at increased speeds. The ignition advance is often provided by spring-controlled centrifugal flyweights. Very often two springs of different stiffness are used to provide two stages of advance rate.

Ignition timing has to be advanced at part throttle settings since the reduced pressure and temperature in the cylinder cause slower combustion. The part throttle condition is defined by the pressure drop between atmosphere and the inlet manifold, the so-called engine 'vacuum'. An exploded view of a typical automotive distributor is shown in figure 4.12. The central shaft is driven at half engine speed (for four-stroke cycles) and the rotor arm directs the HT voltage to the appropriate sparking plug via the distributor cap. For a four-cylinder engine there is a four-lobed cam that operates the contact breaker. The contact breaker and capacitor are mounted on a plate that can rotate a limited amount around the cams relative to the base plate. The position of this plate is controlled by the vacuum unit, a spring-controlled diaphragm that is connected to the inlet manifold. The cams are on a hollow shaft that can rotate around the main shaft. The relative angular position of the two shafts is controlled by the spring-regulated flyweights. The ignition timing is set by rotating the complete distributor relative to the engine. Figure 4.13 shows typical ignition advance curves for engine speed and vacuum.

4.4 Carburettors

The two main types of carburettor discussed here are the fixed jet and the variable jet types. The design of the inlet manifold is as important as the carburettor.

The two main types of carburettor discussed here are the fixed jet and the forms: as vapour, as liquid droplets, and as a liquid film on the manifold walls. The carburettor and manifold have to perform satisfactorily in both steady-state and transient conditions. When an engine is started with a choke (or strangler), fuel floods into the inlet manifold. Under these conditions the engine starts on a very rich mixture and the inlet manifold acts as a surface carburettor; often there will be small ribs to control the flow of liquid fuel.

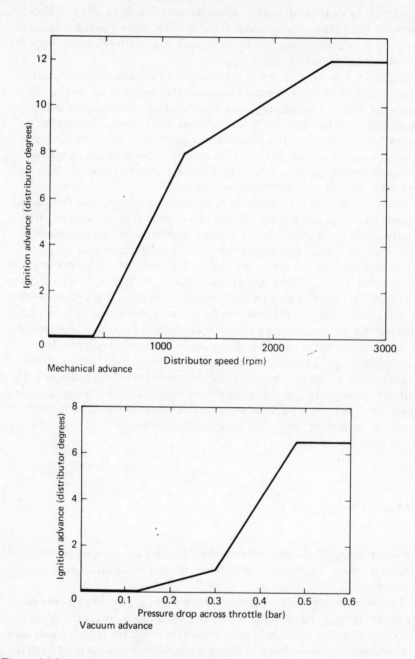

Figure 4.13　　Typical ignition advance curves

In a simple branched manifold the intersections will often have sharp corners, although for good gas flow rounded intersections would be better. The reason is that the sharp corners help to break up the liquid film flowing on the manifold walls, see figure 4.14. In automotive applications the manifold is sometimes inclined relative to the vehicle. This is so that the fuel distribution becomes optimum when the vehicle is ascending a gradient. A way of avoiding the problems in designing the inlet manifold is to use multiple carburettor installations. The problem then becomes one of balancing the carburettors — that is, ensuring that the flow is equal through all carburettors, and that each carburettor is producing the same mixture strength. A cheaper alternative to twin carburettors is the twin choke carburettor. The term 'choke' is slightly misleading, as in this context it means the venturi. The saving in a twin choke carburettor is because there is only one float chamber.

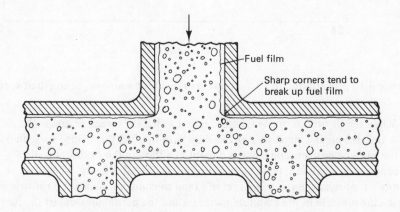

Figure 4.14   Flow of air, fuel vapour, droplets and liquid film in the inlet manifold

Referring back to figure 4.2, points A and B represent the same power output but it is obviously more economical to operate the engine with a wider throttle opening and leaner mixture. In normal operation the carburettor should provide a lean mixture and at full throttle a rich mixture. This ensures economical operation, yet maximum power at full throttle. If a lean mixture were used at full throttle, this would reduce the power output and possibly overheat the exhaust valve because of the slower combustion. When the engine is idling or operating at low load the low pressure in the inlet manifold increases the exhaust gas residuals in the cylinder, and consequently the carburettor has to provide a rich mixture. The way that the optimum air/fuel ratio changes for maximum power and maximum economy with varying power output for a particular engine at constant speed is shown in figure 4.15. The variations in the lean limit and rich limit are also shown.

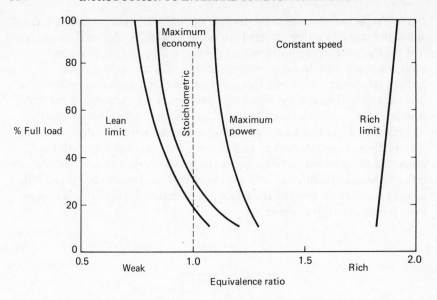

Figure 4.15   Variation in air/fuel ratio requirement with load at constant speed

When the throttle is opened, extra fuel is needed for several reasons. The air flow into the engine increases more rapidly than the fuel flow, since some fuel is in the form of droplets and some is present as a film on the manifold walls. Secondly, for maximum power a rich mixture is needed. Finally, when the throttle is opened the vaporised fuel will tend to condense. When the throttle opens the pressure in the manifold increases and the partial pressure of the fuel vapour will increase (the partial pressure of fuel vapour depends on the air/fuel ratio). If the partial pressure of the fuel rises above its saturation pressure then fuel will condense, and extra fuel is injected to compensate.

Long inlet manifolds will be particularly bad in these respects because of the large volume in the manifold and the length that fuel film and droplets have to travel. In engines with horizontally opposed cylinders it is very difficult to arrange satisfactory carburation from a single carburettor.

When the throttle is suddenly closed, the reduced manifold pressure causes the fuel film to evaporate. This can provide an over-rich mixture, and so lead to emissions of unburnt hydrocarbons. The problem is overcome by a spring-loaded over-run valve on the throttle valve plate that by-passes air into the manifold. Sometimes heated manifolds are used to reduce the liquid film and droplets. The manifold can be heated by the engine coolant, or by conduction from the exhaust manifold. The disadvantage of a heated inlet manifold is the ensuing reduction in volumetric efficiency.

Despite the careful attention paid to manifold design, it is quite usual to have a ±5 per cent variation in mixture strength between cylinders, even for steady-state operation.

4.4.1 Variable jet carburettor

A cross-section of a variable jet or variable venturi carburettor is shown in figure 4.16. The fuel is supplied to the jet ① from an integral float chamber. This has a float-operated valve that maintains a fuel level just below the level of the jet. The pressure downstream of the piston ② is in constant communication with the suction disc (the upper part of the piston) through the passage ③ . If the throttle ④ is opened, the air flow through the venturi ⑤ increases. This decreases the pressure downstream of the venturi and causes the piston ② to rise. The piston will rise until the pressure on the piston is balanced by its weight and the force from the light spring ⑥ . The position of the tapered needle ⑦ in the jet or orifice ① varies with piston position, thus controlling the air/fuel mixture. The damper ⑧ in the oil ⑨ stops the piston ① oscillating when there is a change in load. A valve in the damper causes a stronger damping action when the piston rises than when it falls. When the throttle is opened the piston movement is delayed by the damper, and this causes fuel enrichment of the mixture. For an incompressible fluid the flow through an orifice or venturi is proportional to the square root of the pressure drop. Thus if both air and fuel were incom-

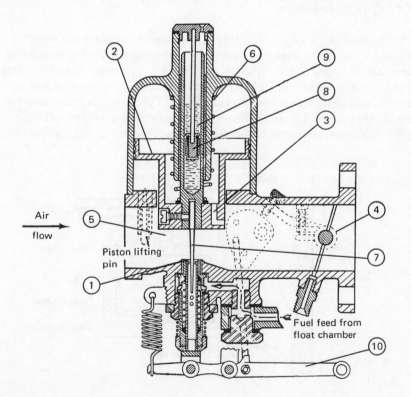

Figure 4.16   Variable jet or variable venturi carburettor (from Judge (1970))

pressible, the air/fuel mixture would be unchanged with increasing flow and a fixed piston. However, as air is compressible its pressure drop will be greater than that predicted by incompressible flow and this will cause extra fuel to flow.

For starting, extra fuel is provided by a lever ⑩ that lowers the jet. A linkage and cam also operate the throttle valve to raise the idling speed. Several modifications are possible to improve the carburettor performance. The position of the jet can be controlled by a bi-metallic strip to allow for the change in fuel properties with temperature. Fuel flow will vary with the eccentricity of the needle in the jet, with the largest flow occurring when the needle touches the side of the jet. Rather than maintain exact concentricity, the needle is lightly sprung so as always to be in contact with the jet.

This type of carburettor should not be confused with carburettors in which there is no separate throttle and the piston and needle are lifted directly. This simple type of carburettor is found on some small engines (such as motorcycles and outboard motors) and does not have facilities like enrichment for acceleration.

4.4.2 Fixed jet carburettor

The cross-section of a simple fixed jet or fixed venturi carburettor is shown in figure 4.17. This carburettor can only sense air flow rate without distinguishing between fully open throttle at a slow engine speed or partially closed throttle at a higher engine speed. The fuel outlet is at the smallest cross-sectional area so that the maximum velocity promotes break-up of the liquid jet and mixing with the air; the minimum pressure also promotes fuel evaporation. The fuel outlet is a few millimetres above the fuel level in the float chamber so that fuel does not

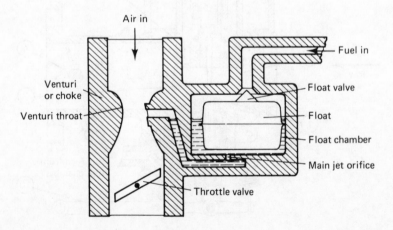

Figure 4.17   Simple fixed jet carburettor

spill or syphon from the float chamber. If the air flow were reversible, there would be no pressure drop; in practice the pressure drop might be 0.05 bar at the maximum air flow rate.

The fuel/air ratio change in response to a change in air flow rate is shown in figure 4.18 for this carburettor. No fuel will flow until the pressure drop in the venturi overcomes the surface tension at the fuel outlet and the head difference from the float chamber. As the air flow increases to its maximum, the air/fuel mixture becomes richer. The maximum air flow is when the velocity at the venturi throat is supersonic. The reason for the change in air/fuel ratio is as follows.

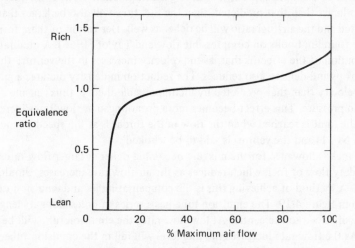

Figure 4.18   Mixture characteristics of a simple carburettor

Fuel can be treated as incompressible, and for flow through an orifice

$$\dot{m}_f = A_o C_o \sqrt{(2\,\rho_f\,\Delta p)} \tag{4.2}$$

where $\dot{m}_f$ = mass flow rate of fuel
          A_o = orifice area
          C_o = coefficient of discharge for the orifice
          ρ_f = density of fuel
          Δp = pressure difference across the orifice.

In contrast, air is compressible, and for flow through the venturi

$$\dot{m}_a = A_t C_v \sqrt{(2\,\rho_a\,\Delta p)} \left[(r)^{1/\gamma} \sqrt{\left\{ \frac{\gamma}{\gamma-1}\ \frac{1-r^{(\gamma-1)/\gamma}}{1-r} \right\}} \right] \tag{4.3}$$

where $\dot{m}_a$ = mass flow rate of air
          A_t = area of venturi throat
          C_v = discharge coefficient for the venturi

ρ_a = density of air at entry to the venturi

Δp = pressure drop between entry and the venturi throat

γ = ratio of gas specific heat capacities

$r = 1 - p/p_a$

p_a = pressure of air at entry to the venturi.

Since r is always less than unity the square bracket term in equation (4.3) will always be less than unity. This term accounts for the compressible nature of the flow. Thus, for a given mass flow rate the pressure drop will be greater than that predicted by a simple approach, assuming incompressible flow. If the pressure drop is larger than that predicted, then the fuel flow will also be larger than expected and the air/fuel ratio will be richer as well. Derivation of these formulae can be found in books on compressible flow and Taylor (1966). A qualitative explanation of the effect is that, as the velocity increases in the venturi the pressure drops and density also reduces. The reduction in density dictates a greater flow velocity than that predicted by incompressible theory, thus causing a greater drop in pressure. This effect becomes more pronounced as flow rates increase, until the limit is reached when the flow in the throat is at the speed of sound (Mach No. 1) and the venturi is said to be 'choked'.

To make allowance for the mixture becoming richer at larger flow rates, a secondary flow of fuel, which reduces as the air flow rate increases, should be added. A method of achieving this is the compensating jet and emulsion tube shown in figure 4.19. The emulsion tube has a series of holes along its length, and an air bleed to the centre. At low flow rates the emulsion tube will be full of fuel. As the flow rate increases the fuel level will fall in the emulsion tube, since air is drawn in through the bleed in addition to the fuel through the compensating jet. The fuel level will be lower inside the emulsion tube than outside it, owing to the pressure drop associated with the air flowing through the emulsion tube holes. As air emerges from the emulsion tube it will evaporate the fuel and form a two-phase flow or emulsion. This secondary flow will assist the break-up of the main flow. The cumulative effects of the main and secondary flows are shown in figure 4.20.

A rich mixture for full throttle operation can be provided by a variety of means, by either sensing throttle position or manifold pressure. The mixture can be enriched by an extra jet (the 'power' jet) or the air supply to the emulsion system can be reduced. Alternatively an air bleed controlled by manifold pressure can be used to dilute a normally rich mixture. This might be a spring-loaded valve that closes when the manifold pressure approaches atmospheric pressure at full throttle.

At low air flow rates no fuel flows, so an additional system is required for idling and slow running. Under these conditions the pressure drop in the venturi is too small and advantage is taken of the pressure drop and venturi effect at the throttle valve. A typical arrangement is shown in figure 4.21. Fuel is drawn into the idling fuel line by the low-pressure region around the throttle valve. A series

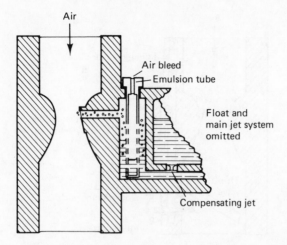

Figure 4.19 Emulsion tube and compensating jet in a fixed jet carburettor

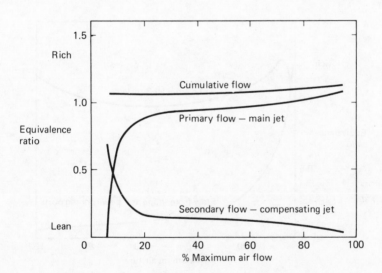

Figure 4.20 Cumulative effect of main jet and compensating jet

of ports are used to provide a smooth progression to the main jet system, and the idling mixture is adjusted by a tapered screw. When this is added to the result of the other jet systems shown in figure 4.20 the results will be as shown in figure 4.22.

With fixed jet carburettors there is no automatic mixture enrichment as the throttle is opened, instead a separate accelerator pump is linked to the throttle. For starting, a rich mixture is provided by a choke or strangler valve at entry to

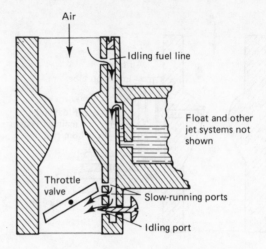

Figure 4.21   Slow-running and idling arrangement in a fixed jet carburettor

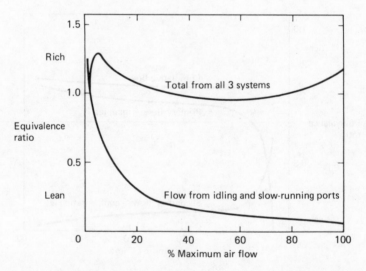

Figure 4.22   Contribution to the air/fuel mixture from the idling and slow-running ports

the carburettor. When this is closed the whole carburettor is below atmospheric pressure and fuel is drawn from the float chamber directly, and the manifold acts as a surface carburettor. The choke valve is spring loaded so that once the engine fires the choke valve is partially opened. The choke is usually linked to a cam that opens the main throttle to raise the idling speed. A complete fixed jet carburettor is shown in figure 4.23; by changing the jet size a carburettor can be adapted for a range of engines.

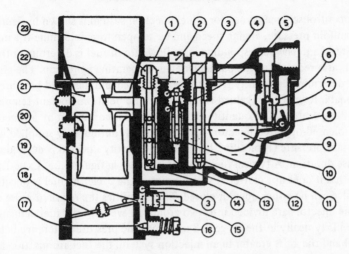

Figure 4.23   Fixed jet or fixed venturi carburettor. ① Air intake; ② idling
jet holder; ③ idling mixture tube; ④ air intake to the bowl;
⑤ air intake for idling mixture; ⑥ needle valve seat; ⑦ needle
valve; ⑧ float fulcrum pivot; ⑨ float; ⑩ carburettor bowl;
⑪ idling jet; ⑫ main jet; ⑬ emulsioning holes; ⑭
emulsioning tube; ⑮ tube for connecting automatic spark
advance; ⑯ idling mixture adjusting screw; ⑰ idling hole to
the throttle chamber; ⑱ throttle butterfly; ⑲ progression
hole; ⑳ choke tube; ㉑ auxiliary venturi; ㉒ discharge
tube; ㉓ emulsioning tube air bleed screw

4.5 Fuel injection

The original purpose of fuel injection was to obtain the maximum power output
from an engine. The pressure drop in a carburettor impairs the volumetric
efficiency of an engine and reduces its power output. The problems of balancing
multiple carburettors and obtaining even distribution in the inlet manifold are
also avoided with fuel injection. Early fuel injection systems were mechanical
and complex two-dimensional cams have now been superseded by electronic
systems.

   Normal practice is to have low-pressure injectors situated close to each inlet
valve. If the injection were direct into the cylinders, there would be problems of
charge stratification. The injectors would also have to withstand the high temperatures in the cylinder and be resistant to the build-up of combustion deposits.
Multi-injector systems also allow the inlet manifold to be designed for optimum
air flow. Single-point injector systems are also possible, and they can take the
form of an electronically controlled carburettor; these are discussed in the next
section.

The control scheme layout for a fuel injection system is shown in figure 4.24. The manifold pressure, engine speed and air temperature determine the mass flow rate of air. The corresponding mass flow rate of fuel is determined from the air/fuel ratio required at the particular engine-operating condition. The air/fuel ratio is determined primarily from the engine speed and manifold pressure, but additional information is gained from the throttle position, coolant temperature and air temperature. The quantity of fuel injected is controlled by a timed pulse to the solenoid-operated injector. Each injector is continuously supplied with fuel at constant pressure (typically 2 bar) and the quantity injected is dependent on the pulse duration. A typical fuel injector is shown in figure 4.25. Experience has shown that it does not matter at which point in the cycle the fuel is injected; the inlet valve can be either open or closed. This simplifies the required pulse out-puts; the injectors are divided into two groups to even out the fuel demand. Pulse duration is typically in the range 2–8 ms. Since the pressure difference between the fuel and the air is greater in an injection system, the fuel atomisation and mixing is better than in a carburettor. The fuel jet can be directed upstream or downstream, but it is often directed on to the inlet valve.

The temperature sensors identify cold start conditions and determine the extra fuel needed, possibly using an extra injector. The throttle position trans-ducer identifies when the throttle is opened and thus controls the extra fuel for acceleration. When the throttle is closed the fuel supply can be stopped until the engine speed reduces to a pre-determined level, say 1000 rpm. Electronic fuel injection can also make allowance for exhaust gas recirculation (EGR) when it is used to reduce the emissions of NO_x.

Figure 4.26 shows a Ford single-point fuel injection system; this is a cheaper alternative to multi-point fuel injection systems, and is widely used in the USA. Again, the higher injection pressure ensures better atomisation than can occur in a carburettor.

The advantage of these digital systems is that they can calculate the ideal amount of fuel during each cycle, rather than using averaged values. Further developments of fuel injection coupled with ignition control are discussed in the next section.

4.6 Electronic control of engines

There are two approaches to electronic control of engines or engine management. The first is to use a memory for storing the optimum values of variables, such as ignition timing and mixture strength, for a set of discrete engine-operating con-ditions. The second approach is to use an adaptive or self-tuning control system to continuously optimise the engine at each operating point. It is also possible to combine the two approaches.

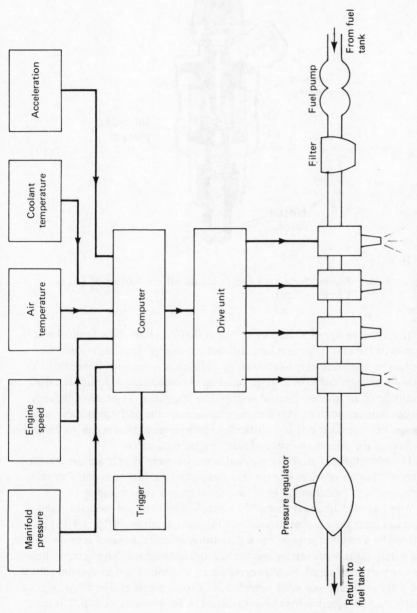

Figure 4.24　Electronic fuel injection system

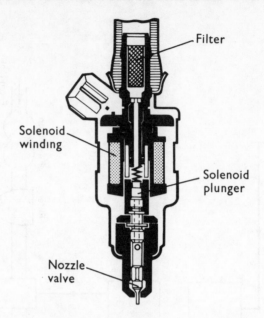

Figure 4.25   Solenoid-operated fuel injection valve (courtesy of Lucas
Electrical Ltd)

The disadvantage of a memory system is that it cannot allow for different
engines of the same type that have different optimum operating conditions
because of manufacturing tolerances. In addition, a memory system cannot
allow for changes due to wear or the build-up of combustion deposits. The disadvantage of an adaptive control system is its complexity. Instead of defining
the operating conditions, it is necessary to measure the performance of the
engine. Furthermore, it is very difficult to provide an optimum control algorithm,
because of the interdependence of many engine parameters.

The advantages of an electronic engine management system are the greater
control it has on variables like ignition timing and mixture strength. The gains
are manifest as reductions in both fuel consumption and emissions.

When ignition timing is controlled mechanically there are inevitably compromises since the vacuum advance and mechanical advance (figure 4.13) are
defined by a series of straight lines. Also the mechanical advance is optimised for
one particular throttle setting and the vacuum advance will have been optimised
for one particular speed. Finally, every possible combination of vacuum advance
and mechanical advance must provide knock-free operation. The advantages of
electronic fuel injection have been discussed in the previous section. The main
advantage is that the air/fuel mixture is based on engine-operating conditions
while carburettors rely principally on air flow rate. With electronic ignition and
fuel injection, combining the electronic control is a logical step since the

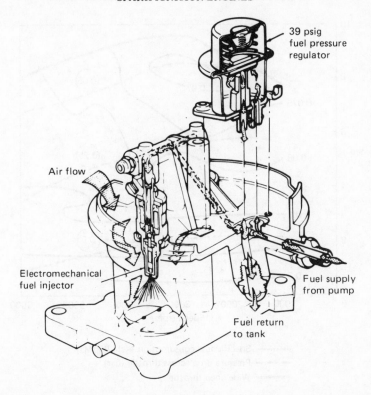

39 psig
fuel pressure
regulator

Air flow

Electromechanical
fuel injector

Fuel supply
from pump

Fuel return
to tank

Figure 4.26   Ford single-point fuel injection system (from Ford (1982))

additional computing power is very cheap. In vehicular applications the natural extension will be control of transmission ratios in order to optimise the overall fuel economy.

With memory systems the engine-operating conditions are derived from engine maps. Figure 4.27 is an example of a typical engine map, derived from experimental results. The map shows contours for fuel economy and manifold pressure. Additional contours could be added for emissions, ignition timing and mixture strength, but these have been omitted to avoid confusion.

When an engine is tested the power output, emissions, manifold depression, optimum ignition timing and air/fuel mixture will all be recorded for each throttle setting and speed. The results are plotted against engine speed and bmep, since bmep is a measure of engine output, independent of its size. In a microprocessor-controlled system the optimum operating conditions will be stored in ROM (Read Only Memory) for each operating point. Since it is difficult to measure the output of an engine, except on a test bed, the operating point is identified by engine speed and manifold pressure.

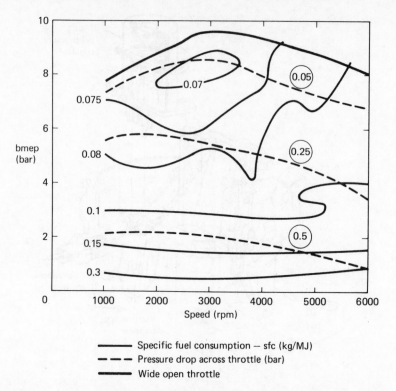

Figure 4.27   Engine map for fuel economy and manifold pressure (reprinted with permission, © 1982 Society of Automotive Engineers, Inc.)

In an adaptive control system the operating point would be found by optimising the fuel economy or emissions. With the multitude of control loops the hierarchy has to be carefully defined, otherwise one loop might be working directly against another. The computational requirements are not a problem since they can be readily met by current microprocessors. Only this approach offers fully flexible systems. For example, ignition timing can be retarded at low engine speeds for ease of starting, and then advanced more than normal for better performance when the engine reaches its normal operating temperature.

Perhaps the best compromise is a memory system, supplemented by adaptive control. For instance, in countries with very strict controls on emissions it is necessary to use a catalyst system and to operate with a stoichiometric air/fuel mixture. By employing an oxygen sensor in the exhaust, the air/fuel mixture can be more closely controlled. In countries with less strict emission controls, high compression ratio engines burning lean air/fuel mixtures will offer the best economy. However, there is a very small operating window (figure 4.7) between the lean combustion limit and the knock limit. If the mixture is too lean the engine will misfire, and this can be detected by fluctuations in the crankshaft speed. The incipience of knock can be detected by sensing chemical species or vibrations before an operator would be aware of the knock.

An important method of improving part load fuel economy is cylinder disablement. In its simplest form this consists of not supplying a cylinder with fuel — a technique used on gas engines at the turn of the century. When applied to fuel-injected multi-cylinder engines a group of cylinders can be disabled, such as a bank of three cylinders in a V6. Alternatively, a varying disablement can be used; for example, every third injection pulse might be omitted in a four-cylinder engine. Slightly greater gains would occur if the appropriate inlet valves were not opened since this would save the pumping losses. This approach could also be used with single-point injection systems or carburettors. However, this requires electrically controlled inlet valves.

In vehicular applications a very significant amount of use occurs in short journeys in which the engine does not reach its optimum temperature — the average British journey is about 10 miles. To improve the fuel economy of carburetted engines under these conditions the choke can be controlled by a stepper motor which responds fast enough to prevent the engine stalling at slower than normal idling speeds.

The questions about electronic engine management concern not just economics but also reliability and whether or not an engine is fail safe.

4.7 Conclusions

Spark ignition engines can operate only within a fairly narrow range of mixture strengths, typically within a gravimetric air/fuel ratio range of 10–18:1 (stoichiometric 14.8:1). The mixture strength for maximum power is about 10 per cent rich of stoichiometric, while the mixture strength for maximum economy is about 10 per cent weak of stoichiometric. This can be explained qualitatively by saying that the rich mixture ensures optimum utilisation of the oxygen (but too rich a mixture will lead to unburnt fuel, which will lower the combustion temperatures and pressures). Conversely the lean mixture ensures optimum combustion of the fuel (but too weak a mixture leads to increasingly significant mechanical losses and ultimately to misfiring). The power output and economy at constant speed for a range of throttle settings are conveniently shown by the so-called 'fish-hook' curves, figure 4.2. From these it is self-evident that at part throttle it is always more economical to run an engine with a weak mixture as opposed to a slightly more closed throttle with a richer mixture.

The main combustion chamber requirements are: compactness, sufficient turbulence, minimised quench areas, and short flame travel from the spark plug to the exhaust valve. Needless to say there is no unique solution, a fact demonstrated by the number of different combustion chamber designs. However, the high-turbulence, high compression ratio, lean-burn combustion chamber should be treated as a separate class. This type of combustion chamber is exemplified by the May Fireball combustion chamber. The high turbulence enables lean mixtures to be burnt rapidly, and the compact combustion chamber (to minimise heat transfer) is located around the exhaust valve. All these attributes contribute to knock-free operation, despite the high compression ratios; the compression ratio can be raised from the usual 9:1 to 15:1, so giving a 15 per cent improvement in fuel economy. However, such combustion chambers place a greater demand on manufacturing tolerances, mixture preparation and distribution, and ignition timing.

Mixture preparation is either by carburettor or by fuel injection. There are two main types of carburettor — fixed or variable venturi (or jet) — and both types can be used for single or multi-carburettor installations. Fuel injection provides a much closer control on mixture preparation and injection is invariably at low pressure into the induction passage. When a single-point injection system is used the problems associated with maldistribution in the inlet manifold still occur, but are ameliorated by the better atomisation of the fuel. It is very difficult (maybe even impossible) to design an inlet manifold that gives both good volumetric efficiency and uniform mixture distribution.

With the decreasing cost and increasing power of micro-electronics it is logical to have complete electronic control of ignition timing and mixture preparation. High compression ratio engines need the sophisticated control to avoid both knock and misfiring. The low-emission engines need careful engine control to ensure operation on a stoichiometric mixture, so that the three-way exhaust catalyst can function properly.

4.8 Example

A variable jet carburettor is designed for a pressure drop (Δp) of 0.02 bar and an air/fuel ratio of 15:1. If the throttle is suddenly opened and the pressure drop quadruples, calculate the percentage increases in air flow and fuel, and the new air/fuel ratio. Neglect surface tension and the difference in height between the jet and the liquid level in the float chamber; atmospheric pressure is 1 bar, and $\gamma = 1.4$.
Immediately after the throttle opens there will be no change in either orifice, as the piston and needle will not have moved.
Rewriting equations (4.2) and (4.3) gives

$$\dot{m}_f = k_f \sqrt{(\Delta p)}$$

$$\dot{m}_a = k_a \sqrt{(\Delta p)} \left[(r)^{1/\gamma} \Big/ \left\{ \frac{\gamma}{\gamma - 1} \frac{1 - r^{(\gamma-1/\gamma)}}{\gamma - 1} \right\} \right]$$

where $r = 1 - \Delta p/p$; k_f, k_a are constants.

If Δp quadruples, the increase in $\dot{m}_f$ is

$$\frac{\sqrt{(4\,\Delta p)} - \sqrt{(\Delta p)}}{\sqrt{(\Delta p)}} \times 100 = 100 \text{ per cent}$$

Using suffix 1 to denote conditions before the step change and suffix 2 to denote conditions after

$$r_1 = 1 - \frac{0.02}{1} = 0.98; \quad r_2 = 1 - \frac{0.08}{1} = 0.92$$

$$\dot{m}_{a1} = k_a\sqrt{(0.02)}\,0.98^{(1/1.4)} \Big/ \sqrt{\left(\frac{1.4}{1.4 - 1} \frac{1 - 0.98^{(1.4-1)/1.4}}{1 - 0.98} \right)}$$

$$= k_a\sqrt{(0.02)}\,0.98^{0.714} \Big/ \sqrt{\left(3.5 \frac{1 - 0.98^{0.286}}{1 - 0.98} \right)} = k_a \times 0.140$$

$$\dot{m}_{a2} = k_a\sqrt{(0.08)}\,0.92^{0.714} \Big/ \sqrt{\left(3.5 \frac{1 - 0.92^{0.286}}{1 - 0.92} \right)} = k_a \times 0.271$$

$$\% \text{ increase in } \dot{m}_a = \frac{0.271 - 0.140}{0.140}\,100 = 93.6 \text{ per cent}$$

$$\text{New air fuel ratio} = \frac{15 \times 0.936}{1.00} : 1 = 14.04{:}1$$

4.9 Problems

4.1 Why does the optimum ignition timing change with engine-operating conditions? What are the advantages of electronic ignition with an electronic control system?

4.2 Explain the principal differences between fixed jet and variable jet carburettors. Why does the mixture strength become richer with increasing flow rate in a simple carburettor?

4.3 What are the air/fuel requirements for a spark ignition engine at different operating conditions? How are these needs met by a fixed jet carburettor?

4.4 List the advantages and disadvantages of electronic fuel injection.

4.5 Contrast high-turbulence, high compression ratio combustion chambers with those designed for lower compression ratios.

4.6 Two spark ignition petrol engines having the same swept volume and compression ratio are running at the same speed with wide open throttles. One engine operates on the two-stroke cycle and the other on the four-stroke cycle. State with reasons:

(i) which has the greater power output.
(ii) which has the higher efficiency.

5 Compression Ignition Engines

5.1 Introduction

Satisfactory operation of compression ignition engines depends on proper control of the air motion and fuel injection. The ideal combustion system should have a high output (bmep), high efficiency, rapid combustion, a clean exhaust and be silent. To some extent these are conflicting requirements; for instance, engine output is directly limited by smoke levels. There are two main classes of combustion chamber: those with direct injection (DI) into the main chamber, figure 5.1, and those with indirect injection (ID), figure 5.4, into some form of divided chamber. The fuel injection system cannot be designed in isolation since satisfactory combustion depends on adequate mixing of the fuel and air. Direct injection engines have inherently less air motion than indirect injection engines and, to compensate, high injection pressures (up to 1000 bar) are used with multiple-hole nozzles. Even so, the speed range is more restricted than for indirect injection engines. Injection requirements for indirect injection engines are less demanding; single-hole injectors with pressures of about 300 bar can be used.

There are two types of injector pump for multi-cylinder engines, either in-line or rotary. The rotary pumps are cheaper, but the limited injection pressure makes them more suited to indirect injection engines.

The minimum useful cylinder volume for a compression ignition engine is about 400 cm³, otherwise the surface-to-volume ratio becomes disadvantageous for the normal compression ratios. The combustion process is also slower than in spark ignition engines and the combined effect is that maximum speeds of compression ignition engines are much less than those of spark ignition engines. Since speed cannot be raised, the output of compression ignition engines is most effectively increased by turbocharging. The additional benefits of turbocharging are improvements in fuel economy, and a reduction in the weight per unit output.

The compression ratio of turbocharged engines has to be reduced, in order to restrict the peak cylinder pressure; the compression ratio is typically in the range 12–24:1. The actual value is usually determined by the cold starting requirements, and the compression ratio is often higher than optimum for either economy or power. Another compromise is the fuel injection pattern. For good cold starting the fuel should be injected into the air, although very often it is directed against a combustion chamber wall to improve combustion control.

There are many different combustion chambers designed for different sizes of engine and different speeds, though inevitably there are many similarities. Very often it will be the application that governs the type of engine adopted. For automotive applications a good power-to-weight ratio is needed and some sacrifice to economy is accepted by using a high-speed engine. For marine or large industrial applications size and weight will matter less, and a large slow-running engine can be used with excellent fuel economy.

All combustion chambers should be designed to minimise heat transfer. This does not of itself significantly improve the engine performance, but it will reduce ignition delay. Also, in a turbocharged engine a higher exhaust temperature will enable more work to be extracted by the exhaust turbine. The so-called adiabatic engine, which minimises (not eliminates) heat losses, uses ceramic materials, and will have higher exhaust temperatures.

Another improvement in efficiency is claimed for the injection of water/fuel emulsions; for example, see Katsoulakos (1983). By using an emulsion containing up to 10 per cent water, improvements in economy of 5–8 per cent are reported. Improvements are not universal, and it has been suggested that they occur only in engines in which the air/fuel mixing has not been optimised. For a given quantity of fuel, a fuel emulsion will have greater momentum and this could lead to better air/fuel mixing. An additional mechanism is that when the small drops of water in the fuel droplets evaporate, they do so explosively and break up the fuel droplet. However, the preparation of a fuel emulsion is expensive and it can lead to problems in the fuel injection equipment. If a fuel emulsion made with untreated water is stored, bacterial growth occurs. Fuel emulsions should reduce NO_x emissions since the evaporation and subsequent dissociation of water reduce the peak temperature.

As in spark ignition engines, NO_x emissions can be reduced by exhaust gas recirculation since this lowers the mean cylinder temperature. Alternatively NO_x emissions can be reduced by retarded injection. However this has an adverse effect on output, economy and emissions of unburnt hydrocarbons and smoke.

5.2 Direct injection (DI) systems

Some typical direct injection combustion chambers given by Howarth (1966) are shown in figure 5.1. Despite the variety of shapes, all the combustion chambers are claimed to give equally good performance in terms of fuel economy, power and emissions, when properly developed. This suggests that the shape is less critical than careful design of the air motion and fuel injection. The most impor-

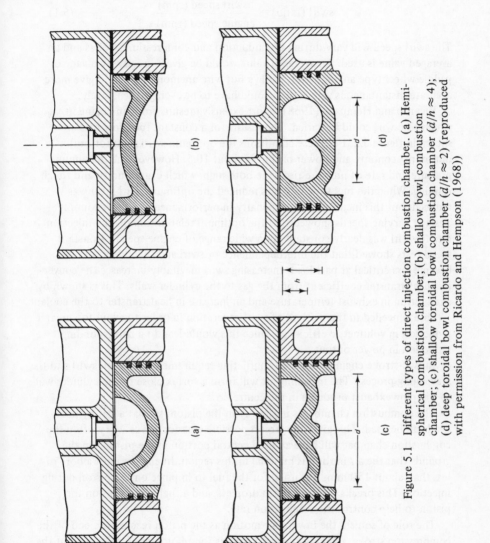

Figure 5.1 Different types of direct injection combustion chamber. (a) Hemispherical combustion chamber; (b) shallow bowl combustion chamber; (c) shallow toroidal bowl combustion chamber ($d/h \approx 4$); (d) deep toroidal bowl combustion chamber ($d/h \approx 2$) (reproduced with permission from Ricardo and Hempson (1968))

tant air motion in direct injection diesel engines is swirl, the ordered rotation of air about the cylinder axis. Swirl can be induced by shrouded or masked inlet valves and by design of the inlet passage — see figure 3.2.

$$\text{swirl (ratio)} = \frac{\text{swirl speed (rpm)}}{\text{engine speed (rpm)}} \qquad (5.1)$$

The swirl speed will vary during the induction and compression strokes and an averaged value is used. An averaged value would be given directly by a vane or paddle-wheel type anemometer, while a hot wire anemometer would give more accurate instantaneous values that would have to be averaged separately.

Ricardo and Hempson (1968) report extensive results from an engine in which the swirl could be varied. The results for a constant fuel flow rate are shown in figure 5.2. For these conditions the optimum swirl ratio for both optimum economy and power output is about 10.5. However, the maximum pressure and rate of pressure rise were both high, which caused noisy and rough running. When the fuel flow rate was reduced the optimum swirl ratio was reduced, but this incurred a slight penalty in performance and fuel economy. Tests at varying speeds showed that the optimum relationship of fuel injection rate to swirl was nearly constant for a wide range of engine speeds. Constant speed tests showed that the interdependence of swirl and fuel injection rate became less critical at part load. Increasing swirl inevitably increases the convective heat transfer coefficient from the gas to the cylinder walls. This is shown by a reduction in exhaust temperature and an increase in heat transfer to the coolant.

Care is needed in the method of swirl generation in order to avoid too great a reduction in volumetric efficiency, since this would lead to a corresponding reduction in power output.

In two-stroke engines there are conflicting requirements between swirl and the scavenging process. The incoming air will form a vortex close to the cylinder wall, so trapping exhaust products in the centre.

The combustion chamber is invariably in the piston, so that a flat cylinder head can be used. This gives the largest possible area for valves. Very often the combustion chamber will have a raised central portion in the piston, on the grounds that the air motion is minimal in this region. In engines with a bore of less than about 150 mm, it is usual for the fuel to impinge on the piston during injection. This breaks up the jet into droplets, and a fuel film forms on the piston to help control the combustion rate.

The role of squish, the inward air motion as the piston reaches the end of the compression stroke, remains unclear. Combustion photography suggests that the turbulence generated by squish does not influence the initial stages of combustion. However, it seems likely that turbulence increases the speed of the final stages of combustion. In coaxial combustion chambers, such as those shown in figure 5.1, squish will increase the swirl rate, by the conservation of angular momentum in the charge.

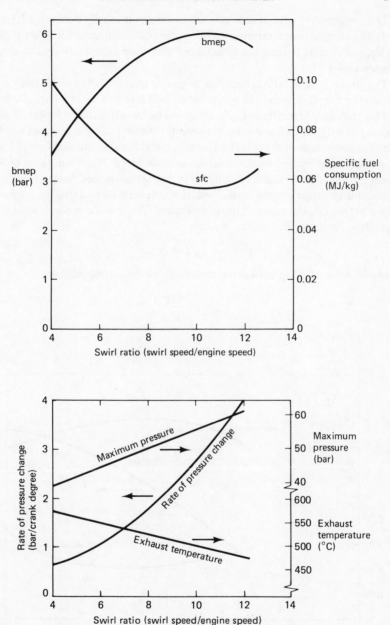

Figure 5.2 Variation of engine performance with swirl ratio at constant fuelling
rate (adapted from Ricardo and Hempson (1968))

The compression ratio of direct injection engines is usually between 12:1 and 16:1. The compact combustion chamber in the piston reduces heat losses from the air, and reliable starting can be achieved with these comparatively low compression ratios.

The stroke-to-bore ratio is likely to be greater than unity (under-square engine) for several reasons. The longer stroke will lead to a more compact combustion chamber. The effect of tolerances will be less critical on a longer stroke engine, since the clearance volume between the piston and cylinder head will be a smaller percentage of the total clearance volume. Finally, the engine speed is most likely to be limited by the acceptable piston speed. Maximum mean piston speeds are about 12 m/s, and this applies to a range of engines, from small automotive units to large marine units. Typical results for a direct injection engine are shown in figure 5.3 in terms of the piston speed. The specific power output is dependent on piston area since

$$\text{power} = \bar{p}_b \, L \, A \, N' \tag{2.13}$$

where N' = no. of firing strokes per second. For a four-stroke engine

$$N' = \frac{n \, v_p}{4L}$$

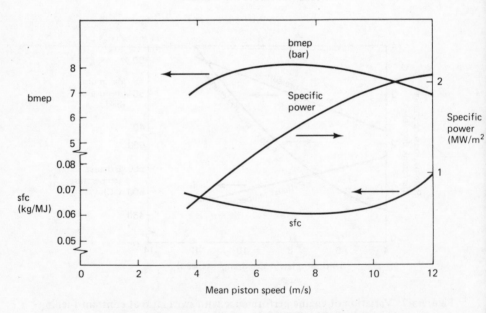

Figure 5.3   Typical performance for direct injection engine (reproduced from Howarth (1966), courtesy of M. H. Howarth, Atlantic Research Associates)

where v_p is the mean piston velocity, and n is the number of cylinders. Thus

$$\text{brake power} = \bar{p}_b \, v_p \, n \, \frac{A}{4} \tag{5.2}$$

But

$$\text{torque} = \frac{\text{power}}{\text{angular velocity}} = \frac{p_b . V}{4\pi}$$

The fuel injector is usually close to the centre-line of the combustion chamber and can be either normal or angled to the cylinder head. To provide good mixing of the fuel and air during the injection period, all the air should move past a jet of fuel. By using multi-hole injectors the amount of swirl can be reduced. However, multi-hole injectors require higher injection pressures for the same flow rate and jet penetration.

In the past the maximum speed of small direct injection engines has been limited by combustion speed. Recently, high-speed direct injection engines have been developed by meticulous attention during the development of the air/fuel mixing; such an engine is discussed in chapter 10.

5.3 Indirect injection (IDI) systems

Indirect injection systems have a divided combustion chamber, with some form of pre-chamber in which the fuel is injected, and a main chamber with the piston and valves. The purpose of a divided combustion chamber is to speed up the combustion process, in order to increase the engine output by increasing engine speed. There are two principal classes of this combustion system; pre-combustion chamber and swirl chamber. Pre-combustion chambers rely on turbulence to increase combustion speed and swirl chambers (which strictly are also pre-combustion chambers) rely on an ordered air motion to raise combustion speed. Howarth (1966) illustrates a range of combustion chambers of both types, see figure 5.4. Both types of combustion chamber use heat-resistant inserts with a low thermal conductivity. The insert is quickly heated up by the combustion process, and then helps to reduce ignition delay. These combustion chambers are much less demanding on the fuel injection equipment. The fuel is injected and impinges on the combustion chamber insert, the jet breaks up and the fuel evaporates. During initial combustion the burning air/fuel mixture is ejected into the main chamber, so generating a lot of turbulence. This ensures rapid combustion in the main chamber without having to provide an ordered air motion during the induction stroke. Since these systems are very effective at mixing air and fuel, a large fraction of the air can be utilised, so giving a high bmep with low emissions

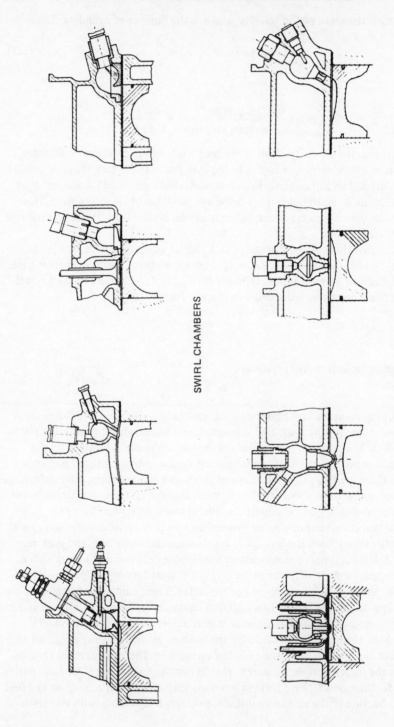

SWIRL CHAMBERS

PRE-COMBUSTION CHAMBERS

of smoke. Unfortunately there are drawbacks. During compression the high gas velocities into the pre-chamber cause high heat transfer coefficients that reduce the air temperature. This means that compression ratios in the range 18–24:1 have to be used to ensure reliable ignition when starting. These compression ratios are higher than optimum for either power output or fuel economy, because of the fall-off in mechanical efficiency. The increased heat transfer also manifests itself as a reduction in efficiency.

Neither type of divided combustion chamber is likely to be applied to two-stroke engines since starting problems would be very acute, and the turbulence generated is likely to interfere with the scavenging process. In a turbocharged two-stroke engine the starting problem would be more acute, but the scavenging problem would be eased.

The fuel injection requirements for both types of divided combustion chamber are less severe, and lower fuel injection pressures are satisfactory. A single orifice in the nozzle is sufficient, but the spray direction should be into the air for good starting and on to the chamber walls for good running. Starting aids like heater plugs are discussed in the next section.

The disadvantages with divided combustion chambers increase in significance as the cylinder size increases. With large cylinders, less advantage can be taken of rapid combustion, and divided combustion chambers are only used in the range of 400–800 cm^3 swept volume per cylinder. By far the most successful combustion chamber for this size range is the Ricardo Comet combustion chamber. The Comet combustion chamber dates back to the 1930s and the current version is the Mk V, see figure 5.5. The volume of the pre-chamber is about half the total clearance volume. The two depressions in the piston induce two vortices of opposing rotation in the gas ejected from the pre-chamber. The insert or 'hot plug' has to be made from a heat-resistant material, since temperatures in the throat can rise to 700°C. Heat transfer from the hot plug to the cylinder head is reduced by minimising the contact area. The temperature of the plug should be sufficient to maintain combustion, otherwise products of partial combustion such as aldehydes would lead to odour in the exhaust.

The direction of fuel injection is critical, see figure 5.5; the first fuel to ignite is furthest from the injector nozzle. This fuel has been in the air longest, and it is also in the hottest air — that which comes into the swirl chamber last. Combustion progresses and the temperature rises; ignition spreads back to within a short distance from the injector. Since the injection is directed downstream of the air swirl, the combustion products are swept away from, and ahead of, the injection path. If the direction of injection is more upstream, the relative velocity between the fuel and the air is greater, so increasing the heat transfer. Consequently the delay period is reduced and cold starting is improved. Unfortunately, the com-

---

Figure 5.4   Different types of pre-combustion and swirl combustion chambers (reproduced from Howarth (1966), courtesy of M. H. Howarth, Atlantic Research Associates)

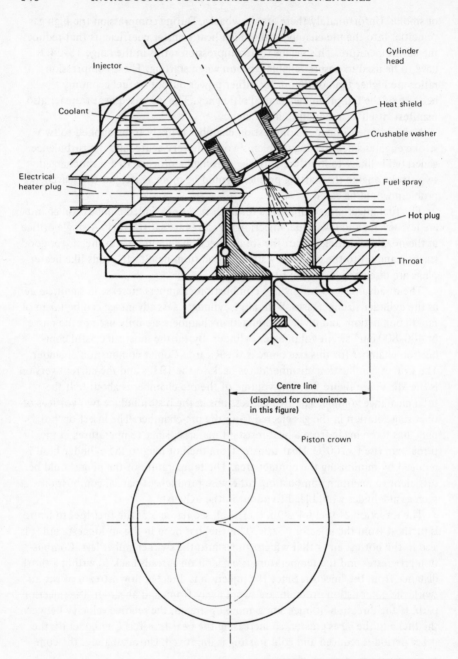

Figure 5.5   Ricardo Comet Mk V combustion chamber (reproduced with permission from Ricardo and Hempson (1968))

bustion products are returned to the combustion zone; this decreases the efficiency and limits the power output.

The high heat transfer coefficients in the swirl chamber can cause problems with injectors; if the temperature rises above 140°C carbonisation of the fuel can occur. The injector temperature can be limited by increasing the cooling or, preferably, by using a heat shield to reduce the heat flow to the injector. Typical performance figures are shown in figure 5.6, which includes part load results at constant speed. As with all compression ignition engines, the power output is limited by the fuelling rate that causes just visible (jv) exhaust smoke. The reduction in economy at part load operation is much less than for a spark ignition engine with its output controlled by throttling.

The Comet combustion system is well suited to engines with twin overhead valves per cylinder. If a four-valve arrangement is chosen then a pre-chamber is perhaps more appropriate.

A final type of divided combustion chamber is the air cell, of which the MAN air cell is an example, see figure 5.7. Fuel is injected into the main chamber and ignites. As combustion proceeds, fuel and air will be forced into the secondary chamber or air cell, so producing turbulence. As the expansion stroke continues the air, fuel and combustion products will flow out of the air cell, so generating further turbulence. In comparison with swirl chambers, starting will be easier since the spray is directed into the main chamber. As the combustion-generated turbulence and swirl will be less, the speed range and performance will be more restricted than in swirl chambers; the air cell is not in common use.

Divided combustion chambers have reduced ignition delay, greater air utilisation, and faster combustion; this permits small engines to run at higher speeds with larger outputs. Alternatively, for a given ignition delay lower quality fuels can be used. As engine size increases the limit on piston speed reduces engine speed, and the ignition delay becomes less significant. Thus, in large engines, direct injection can be used with low-quality fuels. The disadvantage of divided combustion chambers is a 5–15 per cent penalty in fuel economy, and the more complicated combustion chamber design. As energy costs rise the greater economy of direct injection engines has led to the development of small direct injection engines to run at high speeds. The fuel economy of high compression ratio, lean-burn spark ignition engines is comparable with that of indirect injection compression ignition engines.

5.4 Cold starting of compression ignition engines

Starting compression ignition engines from cold is a serious problem. For this reason a compression ratio is often used that is higher than desirable for either optimum economy or power output. None the less, starting can still be a prob-

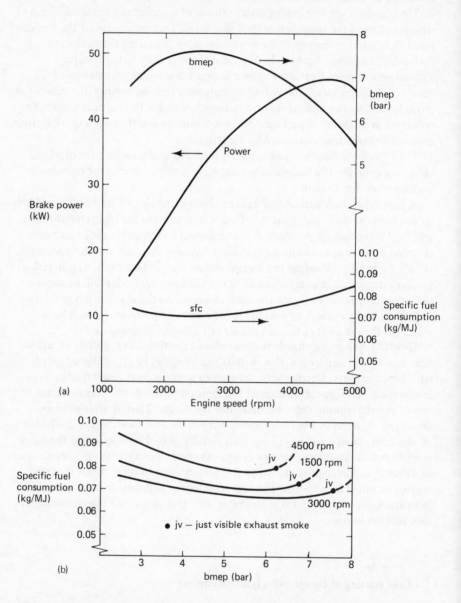

Figure 5.6 Typical performance for a 2 litre engine with Comet combustion
chamber. (a) Full load; (b) part load at different speeds (adapted
from Howarth (1966))

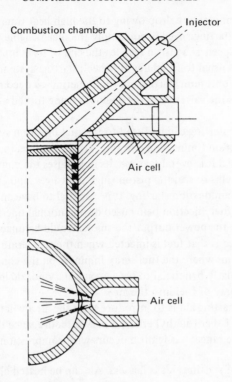

Figure 5.7 MAN air cell combustion chamber

lem, as a result of any of the following: poor-quality fuel, low temperatures, poorly seated valves, leakage past the piston rings or low starting speed. One way of avoiding the compromise in compression ratio would be to use variable compression ratio pistons. In these pistons the distance between the top of the piston (the crown) and the gudgeon pin (little end) can be varied hydraulically. So far these pistons have not been widely used. Ignition in a compression ignition engine relies on both a high temperature and a high pressure. The fuel has to evaporate and reach its self-ignition temperature with sufficient margin to reduce ignition delay to an acceptable level. The high pressure brings air and fuel molecules into more intimate contact, so improving the heat transfer. With too long an ignition delay the expansion stroke will have started before ignition is established.

While at low turning speeds there is greater time for ignition to be established, the peak pressures and temperatures are much reduced. The peak pressure is reduced by greater leakage past the valves and piston rings. The peak temperatures are reduced by the greater time for heat transfer. The situation is worst in divided chamber engines, since the air passes through a cold throat at a high velocity before the fuel is injected. The high velocity causes a small pressure drop, but a large temperature drop owing to the high heat transfer coefficient in the throat. The starting speed is not simply the mean engine speed; of greatest significance is the speed as a piston reaches the end of its compression stroke. For this reason it is usual to fit a large flywheel; starting systems cannot usually turn engines at the optimum starting speed. Sometimes a decompression lever is fitted, which opens the valves until the engine is being turned at a particular starting speed.

Various aids for starting can be fitted to the fuel injection system: excess fuel injection, late injection timing, and extra nozzles in the injector. Excess fuel injection is beneficial for several reasons: its bulk raises the compression ratio, any unburnt fuel helps to seal the piston rings and valves, and extra fuel increases the probability of combustion starting. It is essential to have an interlock to prevent the excess fuel injection being used during normal operation; although this would increase the power output the smoke would be unacceptable. Retarded injection timing means that fuel is injected when the temperature and pressure are higher. In systems where the fuel spray impinges on the combustion chamber surface it is particularly beneficial to have an auxiliary nozzle in the injector, in order to direct a spray of fuel into the air.

An alternative starting aid is to introduce with the air a volatile liquid which self-ignites readily. Ether (diethyl ether) is very effective since it also burns over a very wide mixture range; self-igniton occurs with compression ratios as low as 3.8:1.

The final type of starting aids is heaters. Air can be heated electrically or by a burner prior to induction. A more usual arrangement is to use heater plugs, especially in divided combustion chambers — for example, see figure 5.5. Heater plugs are either exposed loops of thick resistance wire, or finer multi-turned wire insulated by a refractory material and then sheathed. Exposed heater plugs had a low resistance and were often connected in series to a battery. However, a single failure was obviously inconvenient. The sheathed heater plugs can be connected to either 12 or 24 volt supplies through earth and are more robust. With electric starting the heater plugs are switched on prior to cranking the engine. Otherwise the power used in starting would prevent the heater plugs reaching the designed temperature. Ignition occurs when fuel is sprayed on to the surface of the heater; heater plugs do not act by raising the bulk air temperature. With electrical starting systems, low temperatures also reduce the battery performance, so adding further to any starting difficulties. A final possibility would be the use of the high-voltage surface discharge plugs used in gas turbine combustion chambers. Despite a lower energy requirement their use has not been adopted.

In multi-cylinder engines, production tolerances will give rise to variations in compression ratio between the cylinders. However, if one cylinder starts to fire, that is usually sufficient to raise the engine speed sufficiently for the remaining cylinders to fire. In turbocharged engines, the compression ratio is often reduced to limit peak pressures. This obviously has a detrimental effect on the starting performance.

5.5 Fuel injection equipment

A typical fuel injection system is shown in figure 5.8. In general the fuel tank is below the injector pump level, and the lift pump provides a constant-pressure fuel supply (at about 0.75 bar) to the injector pump. The secondary fuel filter contains the pressure-regulating valve, and the fuel bleed also removes any air from the fuel. If air is drawn into the injection pump it cannot provide the correctly metered amount of fuel. It is essential to remove any water or other impurities from the fuel because of the fine clearances in the injection pump and injector. The injection pump contains a governor to control the engine speed. Without a governor the idle speed would vary and the engine could over-speed when the load on the engine is reduced.

The injection pump is directly coupled to the engine (half engine speed for a four-stroke engine) and the pump controls the quantity and timing of the fuel injection (figure 5.9). The quantity of fuel injected will depend on the engine load (figure 5.10). The maximum quantity of fuel that can be injected before the exhaust becomes smoky will vary with speed, and in the case of a turbocharged engine it will also vary with the boost pressure. The injection timing should vary with engine speed, and also load under some circumstances. As the engine speed increases, injection timing should be advanced to allow for the nearly constant ignition delay. In engines that have the injection advance limited by the maximum permissible cylinder pressure, the injection timing can be advanced as the load reduces. Injection timing should be accurate to 1° crank angle.

In multi-cylinder engines equal amounts of fuel should be injected to all cylinders. At maximum load the variation between cylinders should be no more than 3 per cent, otherwise the output of the engine will be limited by the first

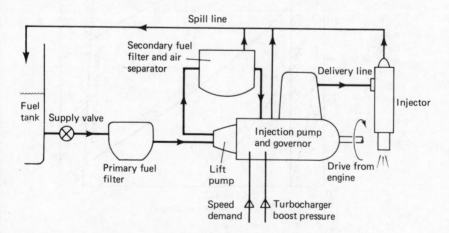

Figure 5.8 Fuel injection system for a compression ignition engine

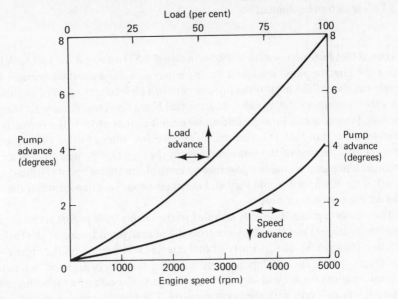

Figure 5.9 Typical fuel injection advance with speed and load

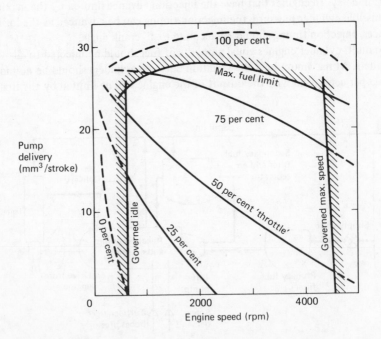

Figure 5.10   Fuel delivery map

cylinder to produce a smoky exhaust. Under idling conditions the inter-cylinder variation can be larger (up to 15 per cent), but the quantities of fuel injected can be as low as 1 mm^3 per cycle. Ricardo and Hempson (1968) and Taylor (1968) provide an introduction to fuel injection equipment, but a much more comprehensive treatment is given by Judge (1967), including detailed descriptions of different manufacturers' equipment. More recent developments include reducing the size of components, but not changing the principles, see Gliken et al. (1979). The matching of fuel injection systems to engines is still largely empirical.

The technology is available for electronic control of injector pumps, and its future application is discussed by Ives and Trenne (1981). Open loop control systems can be used to improve the approximations for pump advance (figure 5.10), but the best results would be obtained with a closed loop control system. However, production of the high injection pressures and the injectors themselves would still have to be mechanical. Electrically operated injectors have never been widely used, since fuel pressures are much greater, up to 1000 bar, and the injection duration is shorter than for spark ignition engines. However, this situation might change as a result of an electrically controlled unit injector developed by Lucas CAV for high-speed direct injection compression ignition engines, see figure 5.11. The unit injector contains both the high-pressure fuel-pumping element, and the injector nozzle. The device is placed in the engine cylinder head, and is driven via a rocker lever from the engine camshaft. The quantity and timing of injection are both controlled electronically through a 'Colenoid' actuator. The 'Colenoid' is a solenoid of patented construction that can respond very quickly (injection periods are typically 1 ms), to control the very high injection pressures (up to 1600 bar).

The electronic control systems could allow for engine speed, load, pressure and exhaust gas recirculation; feedback could also be used to control exhaust emissions.

5.5.2 Fuel injectors

The most important part of the fuel injector is the nozzle; various types of injector nozzle are shown in figure 5.12. All these nozzles have a needle that closes under a spring load when they are not spraying. Open nozzles are used much less than closed nozzles since, although they are less prone to blockage, they dribble. When an injector dribbles, combustion deposits build up on the injector, and the engine exhaust is likely to become smoky. In closed nozzles the needle-opening and needle-closing pressures are determined by the spring load and the projected area of the needle, see figure 5.13. The pressure to open the needle is greater than that required to maintain it open, since in the closed position the projected area of the needle is reduced by the seat contact area. The differential pressures are controlled by the relative needle diameter and seat diameter. A high needle-closing pressure is desirable, since this keeps the nozzle

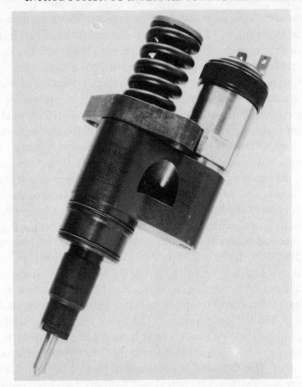

Figure 5.11   Lucas CAV unit injector (courtesy of Lucas CAV)

holes free from blockages caused by combustion deposits. A high needle-closing pressure is also desirable since it maintains a high seat pressure, so giving a better seal. In automotive applications the nozzles are typically about 20 mm in diameter, 45 mm long, with 4 mm diameter needles.

The pintle nozzle, figure 5.14, has a needle or nozzle valve with a pin projecting through the nozzle hole. The shape of the pin controls the spray pattern and the fuel-delivery characteristics. If the pin is stepped, a small quantity of fuel is injected initially and the greater part later. Like all single-hole nozzles the pintle nozzle is less prone to blockage than a multi-hole nozzle. The Pintaux nozzle (PINTle with AUXiliary hole) injector (figure 5.15) was developed by Ricardo and CAV for improved cold starting in indirect injection engines. The spray from the auxiliary hole is directed away from the combustion chamber walls. At the very low speeds when the engine is being started, the delivery rate from the injector pump is low. The pressure rise will lift the needle from its seat, but the delivery rate is low enough to be dissipated through the auxiliary hole without increasing the pressure sufficiently to open the main hole. Once the engine starts, the increased fuel flow rate will cause the needle to lift further, and an increasing amount of fuel flows through the main hole as the engine speed increases.

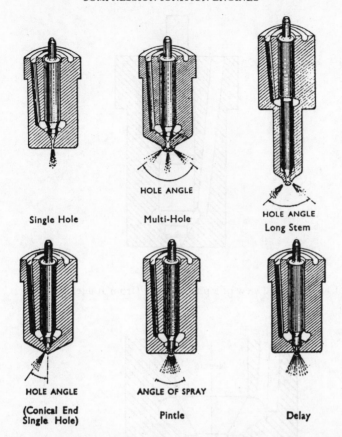

Figure 5.12 Various types of injector nozzle (courtesy of Lucas CAV)

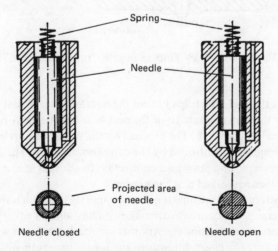

Figure 5.13 Differential action of the injector needle (from Judge (1967))

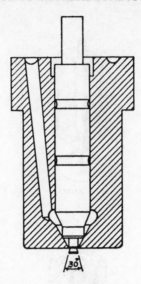

Figure 5.14   Enlarged view of a pintle nozzle (from Judge (1967))

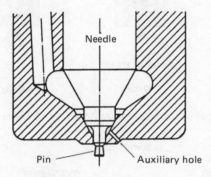

Figure 5.15   The Ricardo–CAV Pintaux nozzle (from Judge (1967))

In all nozzles the fuel flow helps to cool the nozzle. Leakage past the needle is minimised by the very accurate fit of the needle in the nozzle. A complete injector is shown in figure 5.16. The pre-load on the needle or nozzle valve from the compression spring is controlled by the compression screw. The cap nut locks the compression screw, and provides a connection to the spill line, in order to return any fuel that has leaked past the needle.

The spray pattern from the injector is very important, and high-speed combustion photography is very informative. The combustion can either be in an engine or in a special combustion rig. Another approach is to use a water analogue model; here the larger dimensions and longer time scale make observa-

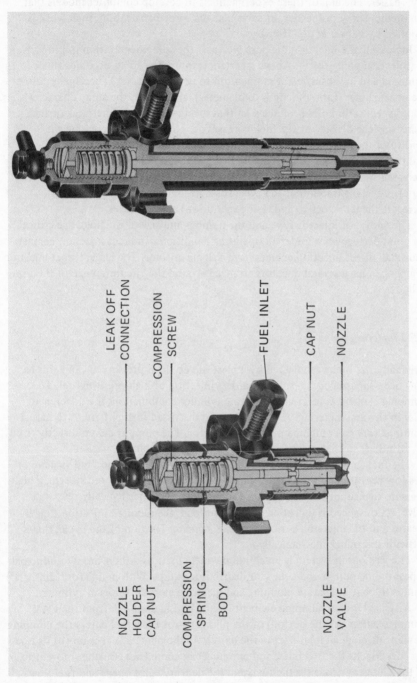

LEAK OFF
CONNECTION

COMPRESSION
SCREW

FUEL INLET

CAP NUT

NOZZLE

NOZZLE
HOLDER
CAP NUT

COMPRESSION
SPRING

BODY

NOZZLE
VALVE

Figure 5.16   Complete fuel injectors (courtesy of Lucas CAV)

tions easier. The aim of these experiments is to develop computer models that can predict spray properties, in particular the spray penetration; such work is reported by Packer *et al.* (1983).

Increasing the injection pressure increases the spray penetration in proportion to (injection pressure)$^{0.4}$. Above a certain pressure the spray becomes finely atomised and has insufficient momentum to penetrate as far. The aspect ratio of the nozzle holes (ratio of length to diameter), also affects the spray characteristics. Long aspect ratio holes produce a jet that diverges less and penetrates further. Increasing the density of the fuel increases penetration, but increasing the density of the air reduces jet penetration. Also, denser fuels are more viscous and this causes the jet to diverge and to atomise less. The jet penetration is also increased with increasing engine speed. The injection period occupies an approximately constant fraction of the cycle for a given load, so as engine speed increases the jet velocity (and thus penetration) also increases.

The choice of injector type and the number and size of the holes are critical for good performance under all operating conditions. Indirect injection engines and small direct injection engines have a single injector. The larger direct injection engines can have several injectors arranged around the circumference of the cylinder.

5.5.2 Injector pumps

Originally the fuel was injected by a blast of very high-pressure air, but this has long been superseded by 'solid' or airless injection of high-pressure fuel. The pumping element is invariably a piston/cylinder combination; the differences arise in the fuel metering. A possibility, not discussed further here, is to have a unit injector − a combined pump and injector. The pump is driven directly from the camshaft so that it is more difficult to vary the timing.

The fuel can be metered at a high pressure, as in the common rail system, or at a low pressure, as in the jerk pump system. In the common rail system, a high-pressure fuel supply to the injector is controlled by a mechanically operated valve. As the speed is increased at constant load, the required injection time increases and the injection period occupies a greater fraction of the cycle; this is difficult to arrange mechanically.

The jerk pump system is much more widely used, and there are two principal types: in-line pumps and rotary or distributor pumps. With in-line (or 'camshaft') pumps there is a separate pumping and metering element for each cylinder.

A typical in-line pumping element is shown in figure 5.17, from the CAV Minimec pump. At the bottom of the plunger stroke (a), fuel enters the pumping element through an inlet port in the barrel. As the plunger moves up (b) its leading edge blocks the fuel inlet, and pumping can commence. Further movement of the plunger pressurises the fuel, and the delivery valve opens and fuel flows towards the injector. The stroke or lift of the plunger is constant and is deter-

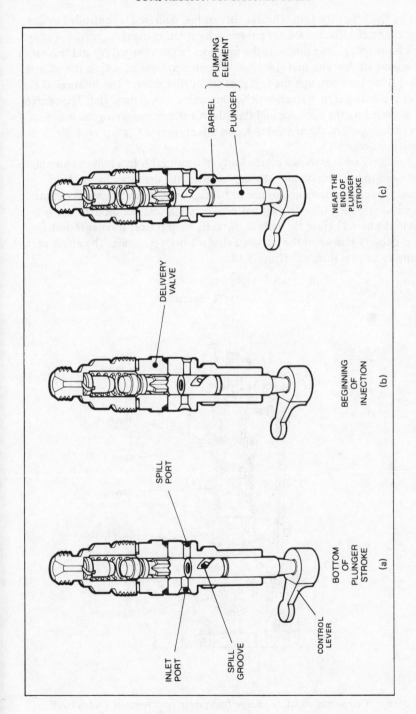

Figure 5.17 Fuel-pumping element from an in-line pump (courtesy of Lucas CAV)

mined by the lift of the cam. The quantity of fuel delivered is controlled by the part of the stroke that is used for pumping. By rotating the plunger, the position at which the spill groove uncovers the spill port can be changed (c) and this varies the pumping stroke. The spill groove is connected to an axial hole in the plunger, and fuel flows back through the spill port to the fuel gallery. The rotation of the plunger is controlled by a lever, which is connected to a control rod. The control rod is actuated by the governor and throttle. An alternative arrangement is to have a rack instead of the control rod, and this engages with gears on each plunger, see figure 5.18.

The delivery valve is shown more clearly in figure 5.17. In addition to acting as a non-return valve it also partially depressurises the delivery pipe to the injector. This enables the injector needle to snap on to its seat, thus preventing the injector dribbling. The delivery valve has a cylindrical section that acts as a piston in the barrel before the valve seats on its conical face; this depressurises the fuel delivery line when the pumping element stops pumping. The effect of the delivery valve is shown in figure 5.19.

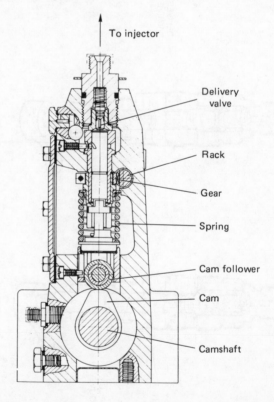

Figure 5.18   Cross-section of an in-line fuel pump (courtesy of Lucas CAV)

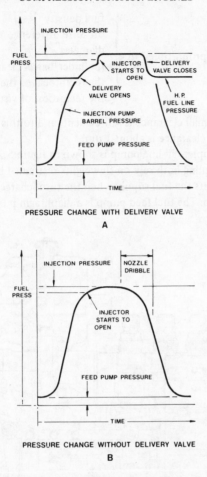

Figure 5.19   Effect of the injector delivery valve (courtesy of Lucas CAV)

Owing to the high pumping pressures the contact stresses on the cam are very high, and a roller type cam follower is used. The fuel pumping is arranged to coincide with the early part of the piston travel while it is accelerating. The spring decelerates the piston at the end of the upstroke, and accelerates the piston at the beginning of the downstroke. The cam profile is carefully designed to avoid the cam follower bouncing.

At high speed the injection time reduces, and the injection pressures will be greater. If accurate fuel metering is to be achieved under all conditions, leakage from the pump element has to be minimal.

leakage is directly proportional to $\begin{cases} \text{fuel density} \\ \text{pressure difference} \\ \text{diameter} \\ \text{(cylinder/barrel clearance)}^3 \\ \text{the reciprocal of the overlap length} \\ \text{the reciprocal of viscosity} \end{cases}$

The importance of a small clearance is self-evident, and to this end the barrel and piston are lapped; the clearance is about 1 μm.

A diagram of a complete in-line pump is shown in figure 5.20. The governor and auto-advance coupling both rely on flyweights restrained by springs. The boost control unit limits the fuel supply when the turbocharger is not at its designed pressure ratio. The fuel feed pump is a diaphragm pump operated off

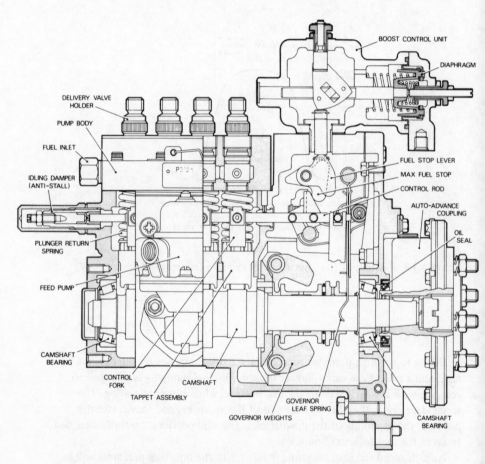

Figure 5.20   Sectional view of the CAV Minimec fuel pump (courtesy of
                Lucas CAV)

the camshaft. In order to equalise the fuel delivery from each pumping element, the position of the control forks can be adjusted on the control rod. The control forks engage on the levers that control the rotation of the plunger.

Rotary or distributor pumps have a single pumping element and a single fuel-metering element. The delivery to the appropriate injector is controlled by a rotor. Such units are more compact and cheaper than an in-line pump with several pumping and metering elements. Calibration problems are avoided, and there are fewer moving parts. However, rotary pumps cannot achieve the same injection pressures as in-line pumps.

The fuel system for a rotary pump is shown in figure 5.21, and figure 5.22 shows the details of the high-pressure pump and rotor for a six-cylinder engine. The transfer pump is a sliding vane pump situated at the end of the rotor. The pump output is proportional to the rotor speed, and the transfer pressure is maintained constant by the regulating valve. The metering value is regulated by the governor, and controls the quantity of fuel that flows to the rotor through the metering port, at the metering pressure. Referring to figure 5.22, the pump plungers that produce the injection pressures rotate in barrels in the rotor. The motion of the plungers comes from a stationary cam with six internal lobes. The phasing is such that the plungers move out when a charging port coincides with the fuel inlet; as the rotor moves round, the fuel is isolated from the inlet. Further rotation causes the distributor port to coincide with an outlet, and the fuel is compressed by the inward movement of the pump plungers.

Governing can be either mechanical or hydraulic, injection timing can be retarded for starting, excess fuel can be provided for cold start, and the turbo-charger boost pressure can be used to regulate the maximum fuel delivery. Injection timing is changed by rotating the cam relative to the rotor. With a vane type pump the output pressure will rise with increasing speed, and this can be used to control the injection advance. A diagram of a complete rotary pump is shown in figure 5.23.

5.5.3 Interconnection of pumps and injectors

The installation of the injector pump and its interconnection with the injectors is critical for satisfactory performance. In a four-stroke engine the injector pump has to be driven at half engine speed. The pump drive has to be stiff so that the pump rotates at a constant speed, despite variations in torque during rotation. If the pump has a compliant drive there will be injection timing errors, and these will be exacerbated if there is also torsional oscillation. The most common drive is either a gear or roller chain system. Reinforced toothed belts can be used for smaller engines, but even so they have to be wider than gear or roller chain drives. The drives usually include some adjustment for the static injection timing.

The behaviour of the complete injection system is influenced by the compressibility effects of the fuel, and the length and size of the interconnecting pipes.

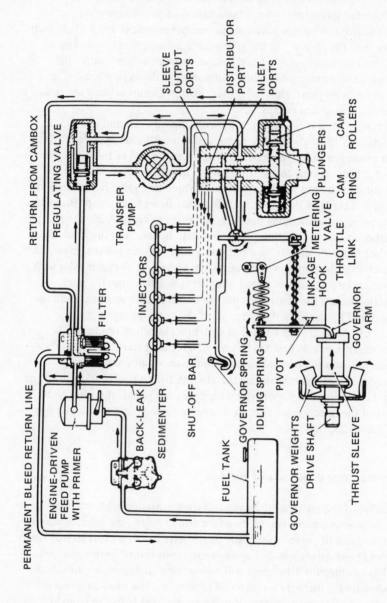

Figure 5.21   Rotary pump fuel system (with acknowledgement to Newton et al. (1983))

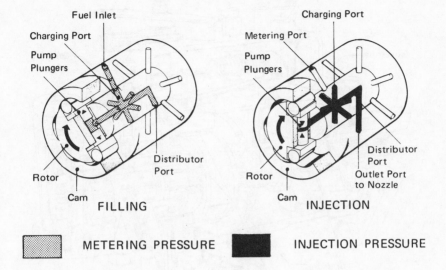

Figure 5.22   Rotor and high-pressure pump from a rotary fuel pump
(courtesy of Lucas CAV)

The compressibility of the fuel is such that an increase in pressure of 180 bar causes a 1 per cent volume reduction. Incidentally, a 1.7 K rise in temperature causes a 1 per cent increase in volume; since the fuel is metered volumetrically this will lead to a reduction in power output. Since the fuel pipes are thick walled, the change in volume is small compared with the effects due to the compressibility of fuel. Pressure (compression or rarefaction) waves are set up between the pump and the injector, and these travel at the speed of sound. The pressure waves cause pressure variations, which can influence the period of injection, the injection pressure and even cause secondary injection.

After the pump delivery valve has opened there will be a delay of about 1 ms per metre of pipe length before the fuel injection begins. To maintain the same fuel injection lag for all cylinders the fuel pipe lengths should be identical. The compression waves may be wholly or partially reflected back at the nozzle as a compression wave if the nozzle is closed, or as a rarefaction wave if the nozzle is open. During a typical injection period of a few milliseconds, waves will travel between the pump and injector several times; viscosity damps these pressure waves. The fuel-line pressure can rise to several times the injection pressure during ejection, and the injection period can be extended by 50 per cent. The volumes of oil in the injector and at the pump have a considerable effect on the pressure waves. Fuel injection systems are prone to several faults, including secondary injection, and after-dribble.

Pressure wave effects can lead to secondary injection — fuel injected after the main injection has finished. Secondary injection can lead to poor fuel consump-

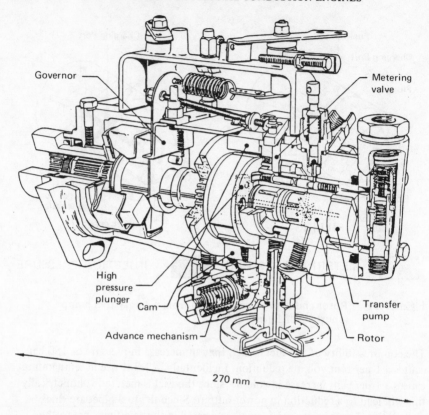

Figure 5.23   Typical rotary or distributor type fuel pump (from Ives and
Trenne (1981))

tion, a smoky exhaust and carbon formation on the injector nozzle. Figure 5.24
shows a fuel-line pressure diagram in which there is a pressure wave after the main
injection period that is sufficient to open the injector. Secondary injection can be
avoided by increasing the fuel-line length, or changing the volumes of fuel at the
pump or injector.

After-dribble is a similar phenomenon; in this case the pressure wave occurs as
the injector should be closing. The injector does not fully close, and some fuel
will enter the nozzle at too low a pressure to form a proper spray.

Problems with interconnecting pipework are of course eliminated with unit
injectors, since the pump and injector are combined. Unit injectors can thus be
found on large engines, since particularly long fuel pipes can be eliminated, and
the bulk of the unit injector can be accommodated more easily. Techniques exist
to predict the full performance of fuel injection systems in order to obtain the
desired injection characteristics — see Knight (1960–61).

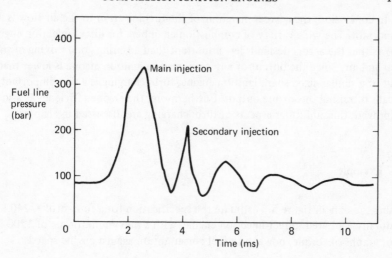

Figure 5.24   Pressure waves in the fuel line from the pump to the injector

5.6 Conclusions

In any type of compression ignition engine it is essential to have properly matched
fuel injection and air motion. These requirements are eased in the case of indirect
injection engines, since the pre-chamber or swirl chamber produces good mixing
of the fuel and air. Since the speed range and the air utilisation are both greater
in the indirect injection engine, the output is greater than in direct injection
engines. However, with divided combustion chambers there is inevitably a
pressure drop and a greater heat transfer, and consequently the efficiency of indirect injection engines is less than the efficiency of direct injection engines. Thus
the development of small high-speed direct injection engines is very significant.

The compression ratio of compression ignition engines is often dictated by
the starting requirements, and it is likely to be higher than optimum for either
maximum fuel economy or power output. This is especially true for indirect
injection engines where the compression ratio is likely to be in the range 18–24:1.
Even so, additional starting aids are often used with compression ignition engines,
notably, excess fuel injection, heaters and special fuels.

Apart from unit injectors there are two main types of injector pump: in-line
pumps and rotary or distributor pumps. Rotary pumps are cheaper, but the
injection pressures are lower than those of in-line pumps. Thus rotary fuel
pumps are better suited to the less-demanding requirements of indirect injection
engines. The fuel injectors and nozzles are also critical components, and like the
injection pumps they are usually made by specialist manufacturers.

In conclusion, the correct matching of the fuel injection to the air flow is all important. The wide variety of combustion chambers for direct injection engines shows that the actual design is less important than ensuring good mixing of the fuel and air. Since the output of any compression ignition engine is lower than that of a similar-sized spark ignition engine, turbocharging is a very important means of raising the engine output. Furthermore, the engine efficiency is also improved; this and other aspects of turbocharging are discussed in chapter 7.

5.7 Example

Using the data in figure 5.3 estimate the specification for a four-stroke, 240 kW, naturally aspirated, direct injection engine, with a maximum torque of 1200 Nm. Plot graphs of torque, power and fuel consumption against engine speed.

First calculate the total required piston area (A_t) assuming a maximum of 2.0 MW/m^2 :

$$A_t = \frac{240 \times 10^3}{2.0 \times 10^6} = 0.12 \text{ m}^2$$

Assuming a maximum bmep $(\bar{p}_b)$ of 8×10^5 N/m^2

$$\text{Power} = T\omega = \bar{p}_b L A_t N^*$$

where $\omega = 2N^*.2\pi$ rad/s, and

$$\text{stroke, } L = \frac{T\omega}{\bar{p}_b A_t N^*} = \frac{1200 \times 4\pi N^*}{8 \times 10^5 \times 0.12 \times N^*} = \underline{0.157 \text{ m}}$$

The number of pistons should be such that the piston diameter is slightly smaller than the stroke. With an initial guess of 8 cylinders:

$$A = \frac{A_t}{8} = \frac{0.12}{8} = 0.015 \text{ m}^2$$

$$\text{bore} = \sqrt{\left(\frac{4 \times 0.015}{\pi}\right)} = \underline{0.138 \text{ m}}$$

a value that would be quite satisfactory for a stroke of 0.157 m.

$$\text{swept volume} = A_t L = 0.12 \times 0.157 = 18.84 \text{ litres}$$

For a maximum mean piston speed (v_p) of 12 m/s, the corresponding engine speed is

$$\frac{v_p \times 60}{2L} = \frac{12 \times 60}{2 \times 0.157} = 2293 \text{ rpm}$$

The results from figure 5.3 can now be replotted as shown in figure 5.25. The final engine specification is in broad agreement with the Rolls Royce CV8 engine. It should be noted that the power output is controlled by the total piston area. Increasing the stroke increases the torque, but will reduce the maximum engine speed, thus giving no gain in power.

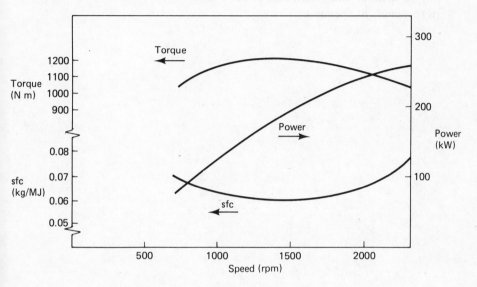

Figure 5.25   Performance curves for engine in example 5.1; 8.8 litre naturally aspirated direct injection engine

5.8 Problems

5.1 Contrast the advantages and disadvantages of indirect and direct injection compression ignition engines.

5.2 Discuss the problems in starting compression ignition engines, and describe the different starting aids.

5.3 Comment on the differences between in-line and rotary fuel injection pumps.

5.4 Describe the different ways of producing controlled air motion in compression ignition engines.

5.5 An engine manufacturer has decided to change one of his engines from a spark ignition type to a compression ignition type. If the swept volume is unchanged, what effect will the change have on:

 (i) maximum torque?
 (ii) maximum power?
(iii) the speed at which maximum power occurs?
(iv) economy of operation?

6 Induction and Exhaust Processes

6.1 Introduction

In reciprocating internal combustion engines the induction and exhaust processes are non-steady flow processes. For many purposes, such as cycle analysis, the flows can be assumed to be steady. This is quite a reasonable assumption, especially for multi-cylinder engines with some form of silencing in the induction and exhaust passages. However, there are many cases for which the flow has to be treated as non-steady, and it is necessary to understand the properties of pulsed flows and how these can interact.

Pulsed flows are very important in the charging and emptying of the combustion chambers, and in the interactions that can occur in the inlet and exhaust manifolds. This is particularly the case for two-stroke engines where there are no separate exhaust and induction strokes. An understanding of pulsed flows is also needed if the optimum performance is to be obtained from a turbocharger; this application is discussed in the next chapter. In naturally aspirated engines it is also important to design the inlet and exhaust manifolds for pulsed flows, if optimum performance and efficiency are to be attained. However, inlet and exhaust manifold designs are often determined by considerations of cost, ease of manufacture, space and ease of assembly, as opposed to optimising the flow.

There is usually some form of silencing on both the inlet and exhaust passages. Again, careful design is needed if large pressure drops are to be avoided.

In four-stroke engines the induction and exhaust processes are controlled by valves. Two-stroke engines do not need separate valves in the combustion chamber, since the flow can be controlled by the piston moving past ports in the cylinder. The different types of valve (poppet, disc, rotary and sleeve) and the different actuating mechanisms are discussed in the next section.

The different types of valve gear, including a historical survey, are described by Smith (1967). The gas flow in the internal combustion engine is covered in some detail by Annand and Roe (1974).

6.2 Valve gear

6.2.1 Valve types

The most commonly used valve is the mushroom-shaped poppet valve. It has the advantage of being cheap, with good flow properties, good seating, easy lubrication and good heat transfer to the cylinder head. Rotary and disc valves are still sometimes used, but are subject to heat transfer, lubrication and clearance problems.

The sleeve valve was once important, particularly for aero-engines prior to the development of the gas turbine. The sleeve valve consisted of a single sleeve or pair of sleeves between the piston and the cylinder, with inlet and exhaust ports. The sleeves were driven at half engine speed and underwent vertical and rotary oscillation. Most development was carried out on engines with a single sleeve valve. There were several advantages associated with sleeve valve engines, and these are pointed out by Ricardo and Hempson (1968), who also give a detailed account of the development work. Sleeve valves eliminated the hot spot associated with a poppet valve. This was very important when only low octane fuels were available, since it permitted the use of higher compression ratios, so leading to higher outputs and greater efficiency. The drive to the sleeve could be at crankshaft level, and this led to a more compact engine when compared with an engine that used overhead poppet valves. The piston lubrication was also improved since there was always relative motion between the piston and the sleeve. In a conventional engine the piston is instantaneously stationary, and this prevents hydrodynamic lubrication at the ends of the piston stroke. This problem is most severe at top dead centre where the pressures and temperatures are greatest.

The disadvantages of the sleeve valve were: the cost and difficulty of manufacture, lubrication and friction between the sleeve and cylinder, and heat transfer from the piston through the sleeve and oil film to the cylinder. Compression ignition engines were also developed with sleeve valves.

6.2.2 Valve-operating systems

In engines with overhead poppet valves (ohv — overhead valves), the camshaft is either mounted in the cylinder block, or in the cylinder head (ohc — overhead camshaft). Figure 6.1 shows an overhead valve engine in which the valves are operated from the camshaft, via cam followers, push rods and rocker arms. This is a cheap solution since the drive to the camshaft is simple (either gear or chain), and the machining is in the cylinder block. In a 'V' engine this arrangement is particularly suitable, since a single camshaft can be mounted in the valley between the two cylinder banks.

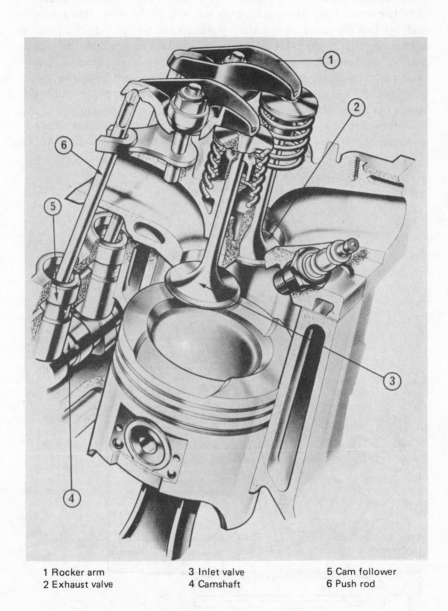

1 Rocker arm   3 Inlet valve   5 Cam follower
2 Exhaust valve   4 Camshaft   6 Push rod

Figure 6.1   Overhead valve engine (courtesy of Ford)

In overhead camshaft (ohc) engines the camshaft can be mounted either directly over the valve stems, as in figure 6.2, or it can be offset. When the camshaft is offset the valves are operated by rockers and the valve clearances can be adjusted by altering the pivot height. The drive to the camshaft is usually by chain or toothed belt. Gear drives are also possible, but would be expensive, noisy and cumbersome with overhead camshafts. The advantage of a toothed belt drive is that it can be mounted externally to the engine, and the rubber damps out torsional vibrations that might otherwise be troublesome.

Referring to figure 6.2, the cam operates on a follower or 'bucket'. The clearance between the follower and the valve end is adjusted by a shim. Although

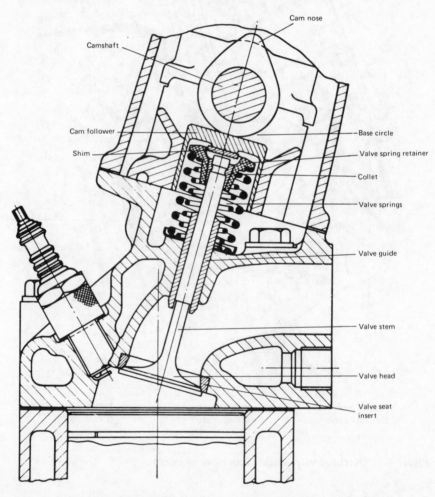

Figure 6.2 Overhead camshaft valve drive (reproduced with permission from Ricardo and Hempson (1968))

this adjustment is more difficult than in systems using rockers, it is much less prone to change. The spring retainer is connected to the valve spindle by a tapered split collet. The valve guide is a press-fit into the cylinder head, so that it can be replaced when worn. Valve seat inserts are used, especially in engines with aluminium alloy cylinder heads, to ensure minimal wear. Normally poppet valves rotate in order to even out any wear, and to maintain good seating. This rotation can be promoted if the centre of the cam is offset from the valve axis. Very often oil seals are placed at the top of the valve guide to restrict the flow of oil into the cylinder. This is most significant with cast iron overhead camshafts which require a copious supply of lubricant.

Not all spark ignition engines have the inlet and exhaust valves in a single line. The notable exceptions are the high-performance engines with hemispherical or pent-roof combustion chambers. Valves in such engines can be operated by various push rod mechanisms, or by twin or double overhead camshafts (dohc). One camshaft operates the inlet valves, and the second camshaft operates the exhaust valves. The disadvantages of this system are the cost of a second camshaft, the more involved machining, and the difficulty of providing an extra drive. An ingenious solution to these problems is the British Leyland 4-valve pent-roof head shown in figure 6.3. A single camshaft operates the inlet valves directly, and the exhaust valves indirectly, through a rocker. Since the same cam lobe is used for the inlet and exhaust valves, the valve phasing is dictated by the valve and rocker geometry. The use of four valves per combustion chamber is quite common in high-performance spark ignition engines, and widely used in the larger compression ignition engines. The advantages of four valves per combustion chamber are: larger valve throat areas for gas flow, smaller valve forces,

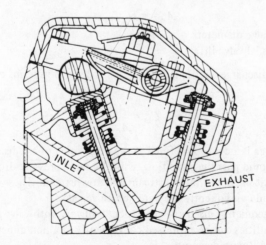

Figure 6.3   Four valve per cylinder pent-roof combustion chamber (from Campbell (1978))

and larger valve seat area. Smaller valve forces occur since a lighter valve with a lighter spring can be used; this will also reduce the hammering effect on the valve seat. The larger valve seat area is important, since this is how heat is transferred (intermittently) from the valve head to the cylinder head.

6.2.3 Flow characteristics of poppet valves

The shape of the valve head and seat are developed empirically to produce the minimum pressure drop. Such experiments are usually carried out on steady-flow air rigs similar to the one shown in figure 6.4. The flow from a fan is decelerated in a diffuser before entering a settling length. To help provide a uniform flow, the type of meshes used in wind tunnels may be useful. The contraction accelerates the flow, thus reducing the relative significance of any turbulence. It is essential for the contraction to match perfectly the inlet passage, otherwise turbulence and extraneous pressure drops will be introduced. The contraction also provides a means of metering the flow rate and of measuring the pressure drop across the valve. The lift setting screw enables the pressure drop to be measured for a range of valve lifts.

A similar arrangement can be used for measuring the flow characteristics of exhaust valves, the exhaust passage being connected to a suction system. Sometimes water rigs are used when the flow patterns are to be investigated. As in all experiments that measure pressure drops through orifices, slightly rounding any sharp corners can have a profound effect on the flow characteristics; care must be taken with all models.

The orifice area can be defined as a 'curtain' area, A_c:

$$A_c = \pi D_v L_v \tag{6.1}$$

where D_v = valve diameter
and L_v = axial valve lift.

This leads to a discharge coefficient (C_D); this is defined in terms of an effective area (A_e):

$$C_D = \frac{A_e}{A_c} \tag{6.2}$$

The effective area is a concept defined as the outlet area of an ideal frictionless nozzle which would pass the same flow, with the same pressure drop, with uniform constant-pressure flows upstream and downstream. These definitions are arbitrary, and consequently they are not universal.

For a given geometry, discharge coefficient will vary with valve lift and flow rate. These quantities can be expressed conveniently as a non-dimensional valve lift (L_v/D_v), and Reynolds number, R_e.

$$R_e = \frac{\rho v x}{\mu} \tag{6.3}$$

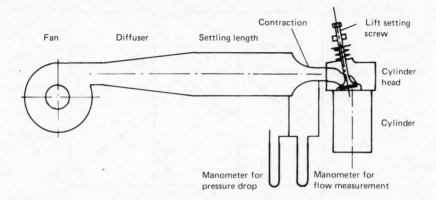

Figure 6.4   Air flow rig to determine the flow characteristics of an inlet valve
(with acknowledgement to Annand and Roe (1974))

where ρ = fluid density
   v = flow velocity
   x = a characteristic length
   μ = fluid viscosity.

The flow characteristics of a sharp-edged inlet valve are shown in figure 6.5.
At low lift the jet fills the gap and adheres to both the valve and the seat. At an
intermediate lift the flow will break away from one of the surfaces, and at high
lifts the jet breaks away from both surfaces to form a free jet. The transition
points will depend on whether the valve is opening or closing. These points are
discussed in detail by Annand and Roe (1974) along with the effects of sharp
corners, radii, and valve seat width. They conclude that a 30° seat angle with a
minimum width seat and 10° angle at the upstream surface gives the best results.
In general, it is advantageous to round all corners on the valve and seat. For
normal valve lifts the effect of Reynolds number on discharge coefficient is
negligible.

The effect of valve lift on discharge coefficient is much smaller for the
exhaust valve — see figure 6.6. The range of pressure ratios across the exhaust
valve is much greater than that across the inlet valve, but the effect on the discharge coefficient is small. The design of the exhaust valve appears less critical
than the inlet valve, and the seat angle should be between 30° and 45°.

In general, the inlet valve is of larger diameter than the exhaust valve, since a
pressure drop during induction has a more detrimental effect on performance
than a pressure drop during the exhaust stroke. For a flat, twin valve cylinder
head, the maximum inlet valve diameter is typically 44–48 per cent and the
maximum exhaust valve diameter is typically 40–44 per cent of the bore
diameter. With pent-roof and hemispherical combustion chambers the valve
sizes can be larger.

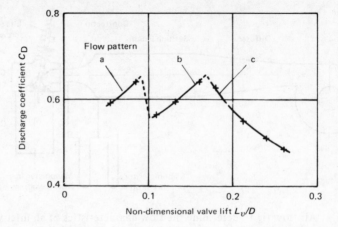

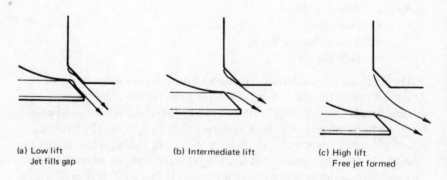

Figure 6.5   Flow characteristics of a sharp-edged inlet valve (with acknowledgement to Annand and Roe (1974))

For a flat four-valve head, as might be used in a compression ignition engine, each inlet valve could be 39 per cent of the bore diameter, and each exhaust valve could be 35 per cent of the bore diameter. This gives about a 60 per cent increase in total valve circumference, or a 30 per cent increase in 'curtain' area (equation 6.1) for the same non-dimensional valve lift. The inlet and exhaust passages should converge slightly to avoid the risk of flow separation with its associated pressure drop. At the inlet side the division between the two valve ports should have a well-rounded nose; this will be insensitive to the angle of incidence of the flow. A knife-edge division wall would be very sensitive to flow breakaway on one side or the other. For the exhaust side the division wall can taper out to a sharp edge.

Annand and Roe (1974) discuss the port arrangements in two-stroke engines.

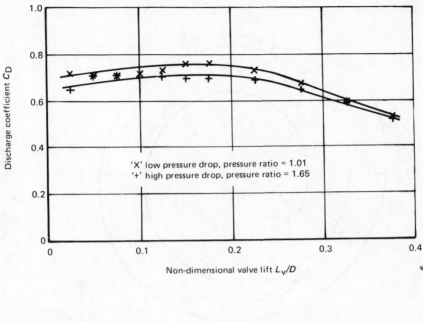

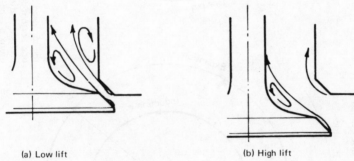

(a) Low lift   (b) High lift

Figure 6.6   Flow characteristics of a sharp-edged 45° seat angle exhaust valve (with acknowledgement to Annand and Roe (1974))

6.2.4 Valve timing

The valve timing dictated by the camshaft and follower system is modified by the dynamic effects that are discussed in the next section. Two timing diagrams are shown in figure 6.7. The first (figure 6.7a) is typical of a compression ignition engine or conventional spark ignition engine, while figure 6.7b is typical of a high-performance spark ignition engine. The greater valve overlap in the second case takes fuller advantage of the pulse effects that can be particularly

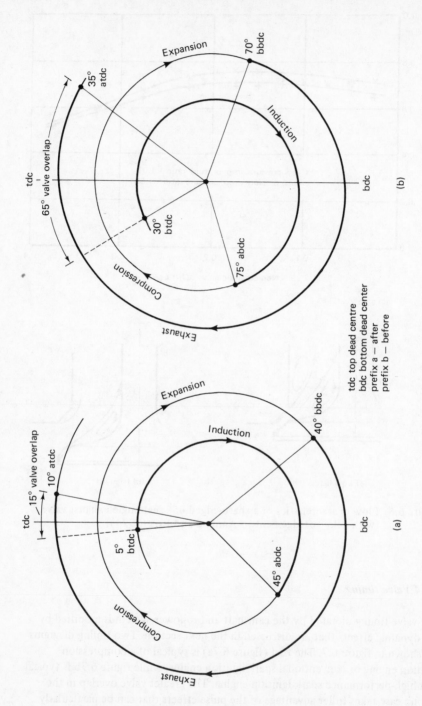

tdc top dead centre
bdc bottom dead center
prefix a – after
prefix b – before

Figure 6.7 Valve-timing diagrams. (a) Small valve overlap; (b) large valve overlap

beneficial at high engine speeds. Turbocharged engines also use large valve overlap periods.

In compression ignition engines the valve overlap at top dead centre is often limited by the piston to cylinder-head clearance. Also the inlet valve has to close soon after bottom dead centre, otherwise the reduction in compression ratio may make cold starting too difficult. The exhaust valve opens about 40° before bottom dead centre (bbdc) in order to ensure that all the combustion products have sufficient time to escape. This entails a slight penalty in the power stroke, but 40° bbdc represents only about 12 per cent of the engine stroke. It should also be remembered that 5° after starting to open the valve may be 1 per cent of fully open, after 10°, 5 per cent of fully open, and not fully open until 120° after starting to open.

In spark ignition engines with large valve overlap, the part throttle and idling operation suffers since the reduced induction manifold pressure causes back-flow of the exhaust. Furthermore, full load economy is poor since some unburnt mixture will pass straight through the engine when both valves are open at top dead centre. These problems are avoided in a turbocharged engine with in-cylinder fuel injection.

In any engine the valve timing will have been optimised for a particular speed and operating condition. In the case of a turbocharged engine the timing may have been optimised in such a way as to make starting difficult. These problems can be overcome by the use of variable valve-timing mechanisms. Parker and Kendrick (1974) review the earlier work and also describe one such camshaft system. A further possibility would be to use early inlet valve closing to regulate the power output of spark ignition engines, thus avoiding the pumping losses associated with throttling. However, none of these mechanisms has yet achieved commercial success.

6.2.5 Dynamic behaviour of valve gear

The theoretical valve motion is defined by the geometry of the cam and its follower. The actual valve motion is modified because of the finite mass and stiffness of the elements in the valve train. These aspects are dealt with after the theoretical valve motion; a fuller discussion is given by Taylor (1968).

The theoretical valve lift, velocity and acceleration are shown in figure 6.8; the lift is the integral of the velocity, and the velocity is the integral of the acceleration. Before the valve starts to move, the clearance has to be taken up. The clearance in the valve drive mechanism ensures that the valve can fully seat under all operating conditions, with sufficient margin to allow for the bedding-in of the valve. To control the impact stresses as the clearance is taken up, the cam is designed to give an initially constant valve velocity. This portion of the cam should be large enough to allow for the different clearance during engine operation. The impact velocity is limited to typically 0.5 m/s at the rated engine speed.

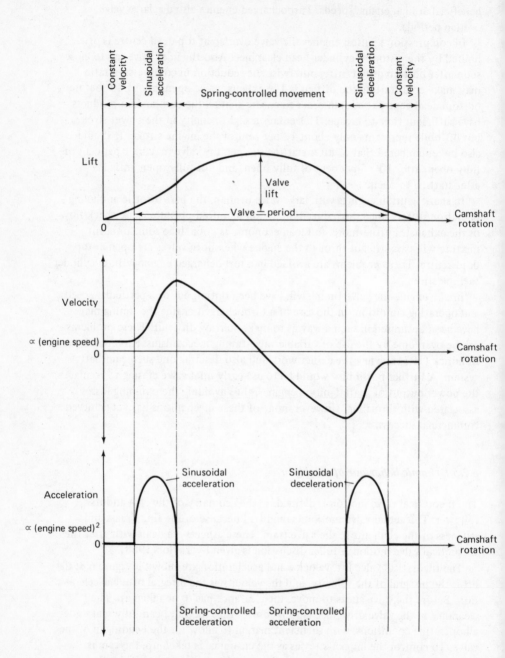

Figure 6.8   Theoretical valve motion

The next stage is when the cam accelerates the valve. The cam could be designed to give a constant acceleration, but this would give rise to shock load-ings, owing to the theoretically instantaneous change of acceleration. A better practice is to use a function that causes the acceleration to rise from zero to a maximum, and then to fall back to zero; both sinusoidal and polynomial functions are appropriate examples. As the valve approaches maximum lift the deceleration is controlled by the valve spring, and as the valve starts to close its acceleration is controlled by the valve spring. The final deceleration is controlled by the cam, and the same considerations apply as before. Finally, the profile of the cam should be such as to give a constant closing velocity, in order to limit the impact stresses.

As can be seen from figure 6.8, the theoretical valve-opening and valve-closing times will depend on the valve clearance. Consequently, the valve timing usually refers to the period between the start of the valve acceleration, and the end of the valve deceleration. The valve lift refers to the lift in the same period, and is usually limited to about $0.25D_v$, to restrict the loads in the valve mechanism.

If the force required during the spring-controlled motion is greater than that provided by the spring, then the valve motion will not follow the cam, and the valve is said to 'jump'. The accelerations will increase in proportion to the square of the engine speed, and a theoretical speed can be calculated at which the valve will jump. The actual speed at which jumping occurs will be below this, because of the elasticity of other components and the finite mass of the spring.

The actual valve motion is modified by the elasticity of the components; a simple model is shown in figure 6.9. A comparison of theoretical and actual valve motion is shown in figure 6.10. Valve bounce can occur if the seating velocity is too great, or if the pre-load from the valve spring is too small. This is likely to lead to failure, especially with the exhaust valve.

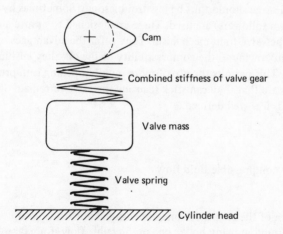

Figure 6.9 Simple valve gear model

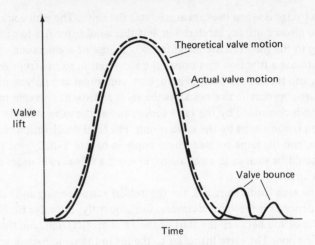

Figure 6.10   Comparison of theoretical and actual valve motion

To minimise the dynamic effects, the valve should be made as light as possible, and the valve gear should be as stiff as possible. The camshaft should have a large diameter shaft with well-supported bearings, and the cams should be as wide as possible. Any intermediate components should be as light and stiff as possible.

More realistic models of valve systems can be used with a computer, which analyses the valve motion and then deduces the corrected cam profiles. The computer can then predict the response for a range of speeds.

Sometimes concentric valve springs are used, especially for engines operating at the highest speeds. An obvious advantage is the reduced tendency for valve jump. Also, springs have a finite mass and are subject to inter-coil vibrations called 'surge'. With two springs the frequencies at which surge occurs will be different, and surge should thus be less troublesome. Sometimes hydraulic tappets (or cam followers) are used. These are designed to ensure a minimum clearance in the valve train mechanism. They offer the advantages of automatic adjustment and, owing to the compressibility of the oil, they cushion the initial valve motion. This permits the use of more rapidly opening cam profiles. The disadvantages are that they can stick (causing the valve to remain open), and the valve motion is less well defined.

6.3 Unsteady compressible fluid flow

The derivation of the results for one-dimensional, unsteady compressible fluid flow can be found in many books on compressible flow, for example Daneshyar (1976); the main results will be quoted and used in this section.

Unsteady flow is treated by considering small disturbances superimposed on a steady flow. For analytical simplicity the flow is treated as adiabatic and reversible, and thus isentropic. The justification for reversibility is that, although the flow may not in fact be frictionless, the disturbances or perturbations are small. Further, the fluid properties are assumed not to change across the perturbation.

By considering the conservation of mass, momentum and energy, the propagation speed for a perturbation or small pressure wave is found to be the speed of sound, a.

For a perfect gas

$$a = \sqrt{\left(\frac{\gamma p}{\rho}\right)} = \sqrt{\gamma_{RT}} \tag{6.4}$$

In a simple wave of finite amplitude, allowance can be made for the change in properties caused by the change in pressure. In particular, an increase in pressure causes an increase in the speed of propagation. A simple wave can be treated by considering it as a series of infinitesimal waves, each of which is isentropic. If the passage of a wave past a point increases the pressure, then it is a compression wave, while if it reduces the pressure then it is an expansion or rarefaction wave – see figure 6.11.

Recalling that an increase in pressure causes an increase in the propagation speed of a wave, then a compression wave will steepen, and an expansion wave will flatten. This is shown in figure 6.12 for a simple wave at four successive times. When any part of the compression wave becomes infinitely steep ($\partial p/\partial x = \infty$) then a small compression shock wave is formed. The shock wave continues to grow and the simple theory is no longer valid, since a shock wave is not isentropic. If the isentropic analysis was valid then the profile shown by the dotted line would form, implying two values of a property (for example, pressure) at a given position and time.

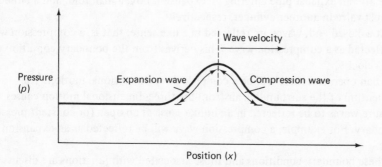

Figure 6.11   Simple pressure wave (adapted from Daneshyar (1976), © 1968 Pergamon Press Ltd)

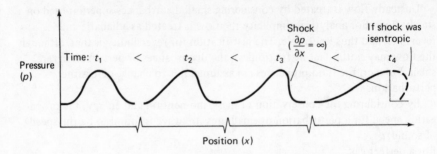

Figure 6.12 Spreading of an expansion wave, steepening of a compression wave (adapted from Daneshyar (1976), © 1968 Pergamon Press Ltd)

The interactions between waves and boundaries can be determined from position diagrams and state diagrams. The state diagrams are not discussed here, but enable the thermodynamic properties, notably pressure, to be determined. The theory and use of state diagrams in conjunction with position diagrams is developed by books such as Daneshyar (1976).

Position diagrams are usually non-dimensional, using the duct length (L_D) to non-dimensionalise position, and the speed of sound (a) and duct length to non-dimensionalise time. The position diagram for two approaching compression waves is shown in figure 6.13; the slopes of the lines correspond to the local value of the speed of sound. The values of the pressure would be obtained from the corresponding state diagram.

For internal combustion engines it is important to know how waves behave at boundaries. There are open ends, junctions and closed ends. Examples of these are: an exhaust pipe entering an expansion box, a manifold, and a closed exhaust valve in another cylinder, respectively.

At a closed end, waves are reflected in a like sense; that is, a compression wave is reflected as a compression wave. This derives from the boundary condition of zero velocity.

At an open end there will be a complex three-dimensional motion. The momentum of the surrounding air from the three-dimensional motion causes pressure waves to be reflected in an unlike sense at an open (or constant-pressure) boundary. For example, a compression wave will be reflected as an expansion wave.

These boundary conditions and those associated with junctions are discussed in detail by Annand and Roe (1974). Although graphical methods give good insight into the mechanisms, computer programs are a more realistic approach to solving pulsating flow problems.

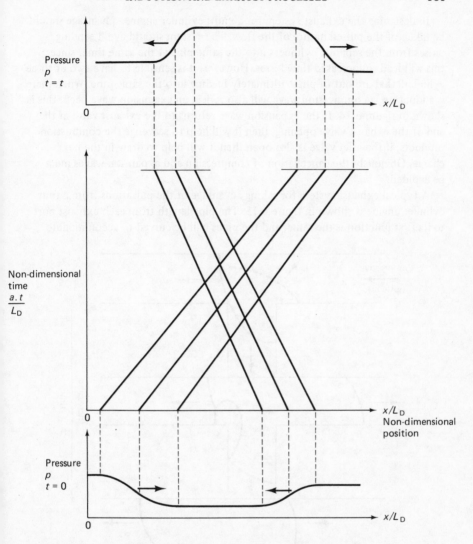

Figure 6.13   Interaction of two approaching compression waves (adapted from Daneshyar (1976), © 1968 Pergamon Press Ltd)

6.4 Manifold design

Most of the comments in this section are aimed at the exhaust system. The pressure pulses in the exhaust system are much greater than those in the inlet system, since in a naturally aspirated engine the pressures in the inlet have to be less than about 1 bar.

In designing the exhaust system for a multi-cylinder engine, advantage should be taken of the pulsed nature of the flow. The system should avoid sending pulses from the separate cylinders into the same pipe at the same time, since this will lead to increased flow losses. However, it is sensible to have two or three cylinders that are out of phase ultimately feeding into the same pipe. When there is a junction, a compression wave will also reflect an expansion wave back; this is shown in figure 6.14. If the expansion wave returns to the exhaust valve at the end of the exhaust valve opening, then it will help to scavenge the combustion products; if the inlet valve is also open then it will help to draw in the next charge. Obviously the cancellation of compression and expansion waves must be avoided.

A typical exhaust system for taking advantage of the pulsations from a four-cylinder engine is shown in figure 6.15. The pipe length from each exhaust port to its first junction is the same, and the pipes will be curved to accommodate

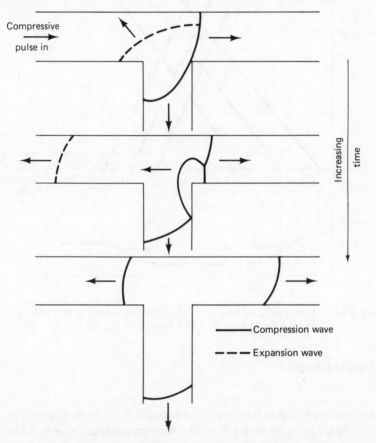

Figure 6.14   Pulsed flow at a junction (with acknowledgement to Annand and Roe (1974))

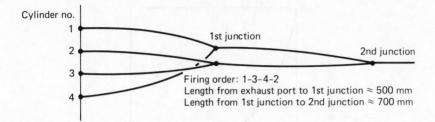

Figure 6.15   Exhaust system for a four-cylinder engine

the specified lengths within the given distances. The length adopted will influence the engine speed at which maximum benefit is obtained. The manifold is such that for the given firing order (cylinders 1–3–4–2), the pressure pulses will be out of phase.

Consider the engine operating with the exhaust valve just opening on cylinder 1. A compression wave will travel to the first junction; since the exhaust valve on cylinder 4 is closed an expansion wave will be reflected back to the open exhaust valve. The same process occurs 180° later in the junction connecting cylinders 2 and 3. At the second junction the flow is significantly steadier and ready for silencing.

Six-cylinder engines use a three-into-one connection at the first junction, with or without a second junction. Eight-cylinder engines can be treated as two groups of four cylinders. If a four-into-one system is used the benefits from pressure pulse interactions occur at much higher engine speeds. The choice of layout always depends on the firing order, and will be influenced by the layout of the engine – whether the cylinders are in-line or in 'V' formation. These points are discussed more fully by Annand and Roe (1974) and Smith (1968).

Induction systems are generally simpler than exhaust systems, especially for engines with fuel injection. The length of the induction pipe will influence the engine speed at which maximum benefit is obtained from the pulsating flow. The lengths shown in figure 6.16 are applicable to engines with fuel injection or a single carburettor per cylinder. In most cases it is impractical to accommodate the ideal length.

Inlet manifolds are usually designed for ease of production and assembly, even on turbocharged engines. When a single carburettor per cylinder is used, the flow pulsations will cause a rich mixture at full throttle as the carburettor will feed fuel for flow in either direction. In engines with a carburettor supplying more than one cylinder the flow at the carburettor will be steadier because of the interaction between compression and expansion waves. The remainder of this section will deal with manifolds for carburetted spark ignition engines.

In carburetted multi-cylinder engines the carburettor is usually connected by a short inlet manifold to the cylinder head. Although a longer inlet passage would have some advantages for a pulsed flow, these would be more than offset

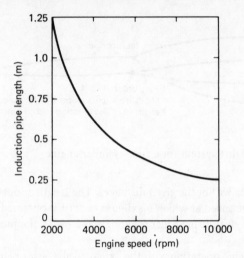

Figure 6.16   Induction pipe length for benefits from pulsating flow (based on Campbell (1978))

by the added delay in response to a change in throttle, caused by fuel lag or 'hold up'. For these reasons it is difficult to devise a central carburettor arrangement for engines with horizontally opposed cylinders.

Even with a four-cylinder in-line engine it is difficult to design a satisfactory single carburettor installation. The manifold shown in figure 6.17a will have a poor volumetric efficiency but can be arranged to give a uniform mixture distribution. In comparison, the manifold in figure 6.17b will have a good volumetric efficiency, but is unlikely to have a uniform mixture distribution. If a twin choke carburettor or two carburettors are fitted to this engine, two of the possible manifold arrangements are shown in figure 6.18. The first arrangement has uniform inlet passages and evenly pulsed flow for the common firing order of 1-3-4-2. The second system (figure 6.18b) is more widely used as it is simpler and equally effective. The pulsed flow will be uneven in each carburettor, so that there will be a tendency for maldistribution with each pair of cylinders. However, as the inlet passages are too short to benefit from any pulse tuning, the effect is not too serious, and is further mitigated by the balance pipe. The balance pipe usually contains an orifice, and the complete geometry has to be optimised by experiment. The same considerations apply to other multi-cylinder engine arrangements. Finally, it should be remembered that the throttle plate can have an adverse effect on the mixture distribution. For example, in figure 6.18a the throttle spindle axes should be parallel to the engine axis.

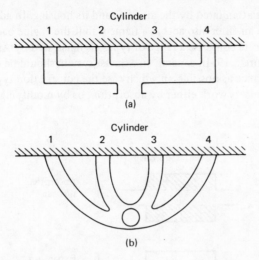

Figure 6.17 Inlet manifolds for single carburettor four-cylinder engine

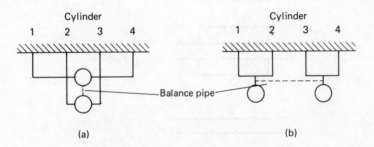

Figure 6.18 Twin carburettor arrangements for a four-cylinder engine

6.5 Silencing

The human ear has a logarithmic response to the magnitude of the fluctuating pressures that are sensed as sound. The ear also has a frequency response, and is most sensitive to frequencies of about 1 kHz. The most effective approach to silencing is the reduction of the peaks, especially those in the most sensitive frequency range of the ear.

The noise from the engine inlet comes from the pulsed nature of the flow, and is modified by the resonating cavities in the cylinder and inlet manifold. A high-frequency hiss is also generated by the vortices being shed from the throttle plate.

The inlet noise is attenuated by the air filter and its housing. In addition to its obvious role, the air filter also acts as a flame trap if the engine back-fires.

Exhaust silencers comprise a range of types, as illustrated by Annand and Roe (1974) — see figure 6.19. In general, an exhaust system should be designed for as low a flow resistance as possible, in which case the constriction type silencer is a poor choice. Silencers work either by absorption, or by modifying the pressure

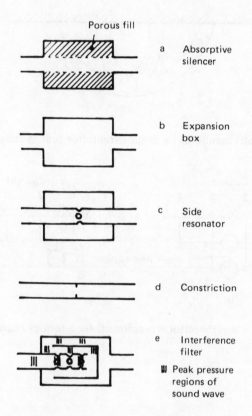

Figure 6.19   The basic silencer elements (with acknowledgement to Annand and Roe (1974))

waves in such a way as to lead to cancellation and a reduction in sound. Absorption silencers work by dissipating the sound energy in a porous medium. Silencers and their connecting pipes should be free of any resonances. Turbochargers tend to absorb the flow pulsations from the engine exhaust, but substitute a high-frequency noise generated by the rotating blades.

6.6 Conclusions

To obtain the optimum performance from any internal combustion engine, great
care is needed in the design of the induction and exhaust systems. Once the type
and disposition of the valve gear have been decided, the valve timing has to be
selected. The ideal valve behaviour is obviously modified by dynamic effects,
owing to the finite mass and elasticity of the valve train components. The actual
valve behaviour can be predicted by computer models. The valve timing will be
determined by the application. The two extremes can be generalised as: normal
spark ignition engines or naturally aspirated compression ignition engines, and
high output spark ignition engines or turbocharged compression ignition engines.
The latter have the greater valve-opening periods. The chosen valve timing is also
influenced by the design of the induction and exhaust passages.

Successful design of the induction and exhaust processes depends on a full
understanding of the pulsed effects in compressible flows. The first solution
methods involved a graphical approach, but these have now been superseded by
computer models. Computer models can also take full account of the flow
variations during the opening and closing of valves, as well as interactions between
the induction and exhaust systems. This approach is obviously very important in
the context of turbochargers, the subject of the next chapter.

6.7 Problems

6.1 Two possible overhead valve combustion chambers are being considered, the
first has two valves and the second design has four valves per cylinder. The
diameter of the inlet valve is 23 mm for the first design and $18\frac{1}{2}$ mm for the
second design. If the second design is adopted, show that the total valve
perimeter is increased by 60.8 per cent. If the valve lift is restricted to the
same fraction of valve diameter, calculate the increase in flow area. What are
the additional benefits in using four valves per cylinder?

6.2 Describe the differences in valve timing on a naturally aspirated diesel engine,
a turbocharged diesel engine, and a high-performance petrol engine.

6.3 Devise an induction and exhaust system for an in-line, six-cylinder, four-
stroke engine with a firing order of 1-5-3-6-2-4, using: (i) twin carburettors,
(ii) triple carburettors.

7 Turbocharging

7.1 Introduction

Turbocharging is a particular form of supercharging in which a compressor is driven by an exhaust gas turbine. The concept of supercharging, supplying pressurised air to an engine, dates back to the beginning of the century. By pressurising the air at inlet to the engine the mass flow rate of air increases, and there can be a corresponding increase in the fuel flow rate. This leads to an increase in power output, and usually an improvement in efficiency since mechanical losses in the engine are not solely dependent on power output. Whether or not there is an improvement in efficiency ultimately depends on the efficiency and matching of the turbocharger or supercharger. Turbocharging does not have a significant effect on exhaust emissions.

Compressors can be divided into two classes: positive displacement and non-positive displacement types. Examples of positive displacement compressors include: Roots, sliding vane, screw, reciprocating piston and Wankel types; some of these are shown in figure 7.1. The axial and radial flow compressors are dynamic or non-positive displacement compressors – see figure 7.2. Because of the nature of the internal flow in dynamic compressors, their rotational speed is an order of magnitude higher than internal combustion engines or positive displacement compressors. Consequently, positive displacement compressors are more readily driven from the engine crankshaft, an arrangement usually referred to as a 'supercharger'. Axial and radial compressors can most appropriately be driven by a turbine, thus forming a turbocharger. Again the turbine can be of an axial or radial flow type. The thermodynamic advantage of turbochargers over superchargers stems from their use of the exhaust gas energy during blow-down, figure 2.5.

A final type of supercharger is the Brown Boveri Comprex pressure wave supercharger shown in figure 7.3. The paddle-wheel type rotor is driven from the engine crankshaft, yet the air is compressed by the pressure waves from the exhaust. Some mixing of the inlet and exhaust gases will occur, but this is not significant.

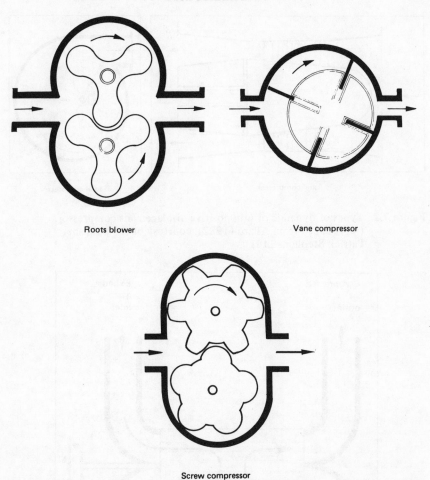

Roots blower Vane compressor

Screw compressor

Figure 7.1 Types of positive displacement compressor (reproduced from Allard (1982), courtesy of the publisher Patrick Stephens Ltd)

The characteristics of turbochargers are fundamentally different from those of reciprocating internal combustion engines, and this leads to complex matching problems when they are combined. The inertia of the rotor also causes a delay in response to changes in load — turbolag. Superchargers have the added complication of a mechanical drive, and the compressor efficiencies are usually such that the overall economy is reduced. However, the flow characteristics are better matched, and the transient response is good because of the direct drive. The Comprex supercharger absorbs minimal power from its drive, and has a good transient response; but it is expensive to make and requires a drive. The fuel economy is worse than a turbocharger, and its thermal loading is higher.

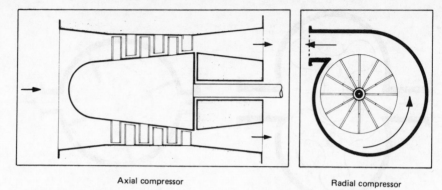

Axial compressor   Radial compressor

Figure 7.2   Types of dynamic or non-positive displacement compressor
(reproduced from Allard (1982), courtesy of the publisher
Patrick Stephens Ltd)

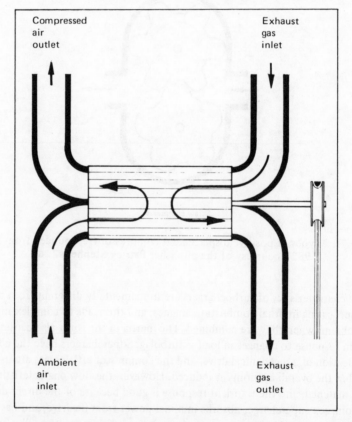

Figure 7.3   Brown Boveri Comprex supercharger (reproduced from Allard
(1982), courtesy of the publisher Patrick Stephens Ltd)

Comprex superchargers have not been widely used, and superchargers are used on spark ignition engines only where the main consideration is power output. Turbochargers have been used for a long time on larger compression ignition engines, and are now being used increasingly on automotive compression ignition and spark ignition engines.

Compound engines are also likely to gain in importance. A compound engine has a turbine geared to the engine crankshaft, either the same turbine that drives the compressor or a separate power turbine. The gearing is usually a differential epicyclic arrangement, and if matching is to be optimised over a range of speeds a variable ratio drive is also needed. Such combinations are discussed by Wallace *et al.* (1983) and by Watson and Janota (1982). Compound engines offer improvements in efficiency of a few per cent compared with conventional turbocharged diesel engines.

Commercial and marketing factors also influence the use of turbochargers. A turbocharged engine will fit in the existing vehicle range, and would not need the new manufacturing facilities associated with a larger engine.

Allard (1982) provides a practical guide to turbocharging and supercharging, and Watson and Janota (1982) give a rigorous treatment of turbocharging. The remainder of this chapter is devoted to turbocharging.

7.2 Radial flow and axial flow machines

The turbomachinery theory applied to turbochargers is the same as for gas turbines, and is covered in books on gas turbines such as Harman (1981), and in books on turbocharging, see Watson and Janota (1982). As well as providing the theory, gas turbines also provided the materials technology for the high temperatures and stresses in turbochargers. Provided that the turbocharger is efficient enough to raise the engine inlet pressure above the exhaust pressure of the engine, the intake and exhaust processes will both benefit. This is particularly significant for engines with in-cylinder fuel injection (since unburnt fuel will not pass straight through the engine), and for two-stroke engines (since there are no separate induction and exhaust strokes).

The efficiency of turbines and compressors depends on their type (axial or radial flow) and size. Efficiency increases with size, since the losses associated with the clearances around the blades become less significant in large machines. These effects are less severe in radial flow machines, so although they are inherently less efficient than axial machines their relative efficiency is better in the smaller sizes.

Compressors are particularly difficult to design since there is always a risk of back-flow, and a tendency for the flow to separate from the blades in the divergent passages. Dynamic compressors work by accelerating the flow in a rotor, giving a rise in total or dynamic pressure, and then decelerating the flow in a diffuser to produce a static pressure rise. Radial compressors are more tolerant of different flow conditions, and they can also achieve pressure ratios of 3.5:1 in a single stage; an axial compressor would require several rotor/stator stages for the same pressure ratio.

A typical automotive turbocharger is shown in figure 7.4, with a radial flow compressor and turbine. For the large turbochargers used in marine applications, the turbine is large enough to be designed more efficiently as an axial flow turbine — see figure 7.5.

The operation of a compressor or turbine is most sensibly shown on a temperature/entropy (T-s) plot. This contrasts with the Otto and Diesel cycles which are conventionally drawn on pressure/volume diagrams. The ideal compressor is both adiabatic and reversible and is thus isentropic — a process represented by a vertical line on the T-s plot, figure 7.6. The suffix s denotes an isentropic process. Real processes are of course irreversible, and are associated with an increase in entropy; this is shown with dotted lines on figure 7.6. Expressions for work (per unit mass flow) can be found using the simplified version of the steady-flow energy equation

$$h_{\text{in}} + Q = h_{\text{out}} + W$$

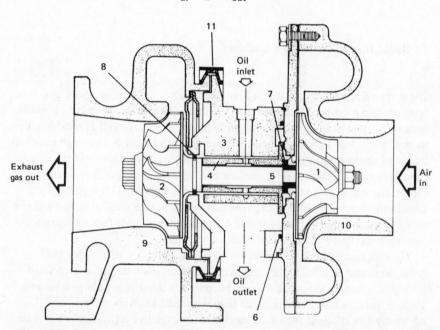

Key: 1 *Compressor wheel.* 2 *Turbine wheel.* 3 *Bearing housing.* 4 *Bearing.* 5 *Shaft.* 6 *seal ('O' ring).* 7 *Mechanical face seal.* 8 *Piston ring seal.* 9 *Turbine housing.* 10 *Compressor housing.* 11 *'V' band clamp.*

Figure 7.4   Automotive turbocharger with radial compressor and radial turbine (reproduced from Allard (1982), courtesy of the publisher Patrick Stephens Ltd)

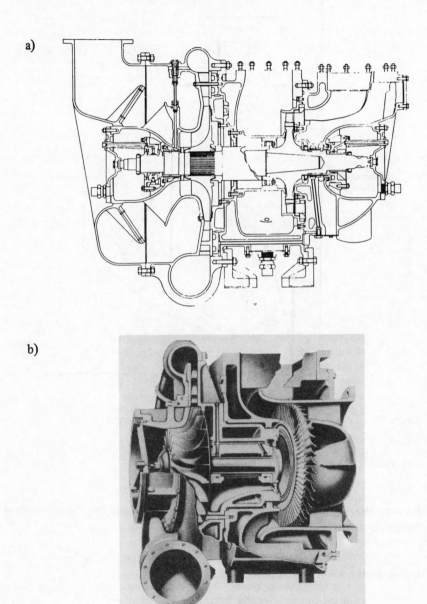

Figure 7.5   Marine turbochargers with radial compressors and axial turbines.
(a) Napier; (b) Elliot (with acknowledgement to Watson and Janota
(1982))

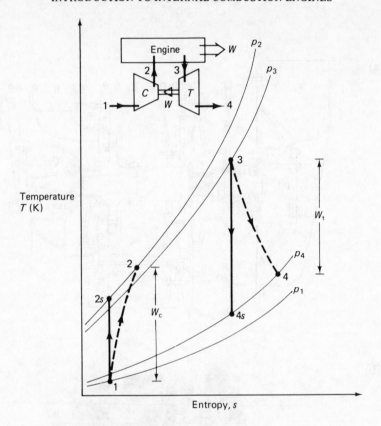

Figure 7.6 Temperature/entropy diagram for a turbocharger

and, since the processes are treated as adiabatic

$$W = h_{\text{in}} - h_{\text{out}}$$

No assumptions about irreversibility have been made in applying the steady-flow energy equation; thus

$$\text{turbine work, } W_{\text{t}} = h_3 - h_4 \qquad (7.1)$$

and defining compressor work as a negative quantity

$$W_{\text{c}} = h_2 - h_1$$

For real gases, enthalpy is a strong function of temperature and a weak function of pressure. For semi-perfect gases, enthalpy is solely a function of temperature, and this is sufficiently accurate for most purposes. Thus

$$W_{\text{c}} = c_{\text{p}} \, (T_2 - T_1) \, \text{kJ/kg} \qquad (7.2)$$

and

$$W_t = c_p (T_3 - T_4) \text{ kJ/ kg} \tag{7.3}$$

where c_p is an appropriate mean value of the specific heat capacity. Consequently the T-s plot gives a direct indication of the relative compressor and turbine works.

This leads to isentropic efficiencies that compare the actual work with the ideal work.

$$\text{compressor isentropic efficiency, } \eta_c = \frac{h_{2s} - h_1}{h_2 - h_1} = \frac{T_{2s} - T_1}{T_2 - T_1}$$

$$\text{and turbine isentropic efficiency, } \eta_t = \frac{h_3 - h_4}{h_3 - h_{4s}} = \frac{T_3 - T_4}{T_3 - T_{4s}}$$

It may appear unrealistic to treat an uninsulated turbine that is incandescent as being adiabatic. However, the heat transferred will still be small compared to the energy flow through the turbine. Strictly speaking, the kinetic energy terms should be included in the steady-flow energy equation. Since the kinetic energy can be incorporated into the enthalpy term, the preceding arguments still apply by using stagnation or total enthalpy.

The shape of the isobars (lines of constant pressure) can be found quite readily. From the 2nd Law of Thermodynamics

$$T \, ds = dh - v dp$$

Thus $\qquad \left(\dfrac{\partial h}{\partial s}\right)_p = T$

or $\qquad \left(\dfrac{\partial T}{\partial s}\right)_p \propto T \qquad$ that is, on the T-s plot isobars have a positive slope proportional to the absolute temperature

and $\qquad \left(\dfrac{\partial h}{\partial p}\right)_s = v \qquad$ that is, the vertical separation between isobars is proportional to the specific volume, and specific volume increases with temperature

Consequently the isobars diverge in the manner shown in figure 7.6.

In a turbocharger the compressor is driven solely by the turbine, and a mechanical efficiency can be defined as

$$\eta_m = \frac{W_c}{W_t} = \frac{m_{12} \, c_{p12} \, (T_2 - T_1)}{m_{34} \, c_{p34} \, (T_3 - T_4)} \tag{7.4}$$

As in gas turbines, the pressure ratios across the compressor and turbine are very important. From the pressure ratio the isentropic temperature ratio can be found:

$$\frac{T_{2s}}{T_1} = \left(\frac{p_2}{p_1}\right)^{(\gamma-1)/\gamma} \quad \text{and} \quad \frac{T_3}{T_{4s}} = \left(\frac{p_3}{p_4}\right)^{(\gamma-1)/\gamma} \tag{7.5}$$

The actual temperatures, T_2 and T_4, can then be found from the respective isentropic efficiencies.

In constant-pressure turbocharging it is desirable for the inlet pressure to be greater than the exhaust pressure $(p_2/p_3 > 1)$, in order to produce good scavenging. This imposes limitations on the overall turbocharger efficiency $(\eta_m \cdot \eta_T \cdot \eta_c)$ for different engine exhaust temperatures (T_3). This is shown in figure 7.7. The analysis for these results originates from the above expressions, and is given by Watson and Janota (1982). Example 7.1 also illustrates the work balance in a turbocharger.

The flow characteristics of an axial and radial compressor are compared in figure 7.8. The isentropic efficiencies would by typical of optimum-sized machines, with the axial compressor being much larger than the radial compressor. Since the turbocharger compressor is very small the actual efficiencies will be lower, especially in the case of an axial machine. The surge line marks the region of unstable operation, with flow reversal etc. The position of the surge line will also be influenced by the installation on the engine. Figure 7.9 shows the wider operating regime of a radial flow compressor. The isentropic efficiency of a turbocharger radial compressor is typically in the range 65–75 per cent.

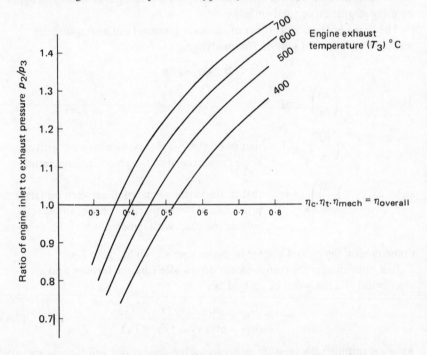

Figure 7.7   Effect of overall turbocharger on the pressure ratio between engine inlet and exhaust manifold pressures, for a 2:1 compressor pressure ratio $(p_2/p_1 = 2)$ with different engine exhaust temperatures (with acknowledgement to Watson and Janota (1982))

The design of turbines is much less sensitive and the isentropic efficiency varies less in the operating range, and rises to over 90 per cent for aircraft gas turbines. The isentropic efficiency of turbocharger turbines is typically 70–85 per cent for radial flow and 80–90 per cent for axial flow machines. These are optimistic 'total to total' efficiencies that assume recovery of the kinetic energy in the turbine exhaust.

A detailed discussion of the internal flow, design, and performance of turbochargers can be found in Watson and Janota (1982).

The flow from an engine is unsteady, owing to the pulses associated with the exhaust from each cylinder, yet turbines are most efficient with a steady flow. If the exhaust flow is smoothed by using a plenum chamber, then some of the energy associated with the pulses is lost. The usual practice is to design a turbine for pulsed flow and to accept the lower turbine efficiency. However, if the compressor pressure ratio is above 3:1 the pressure drop across the turbine becomes excessive for a single stage. Since a multi-stage turbine for pulsed flow is difficult to design at high pressure ratios, a steady constant-pressure turbocharging system should be adopted.

In pulse turbocharging systems the area of the exhaust pipes should be close to the curtain area of the valves at full valve lift. Some of the gain in using small exhaust pipes comes from avoiding the expansion loss at the beginning of blowdown. In addition, the kinetic energy of the gas is preserved until the turbine entry. To reduce frictional losses the pipes should be as short as possible.

For four-stroke engines no more than three cylinders should feed the same turbine inlet. Otherwise there will be interactions between cylinders exhausting at the same time. For a four-cylinder or six-cylinder engine a turbine with two inlets should be used. The exhaust connections should be such as to evenly space the exhaust pulses, and the exhaust pipes should be free of restrictions or sharp corners. Turbines with four separate entries are available, but for large engines it can be more convenient to use two separate turbochargers. For a 12-cylinder engine two turbochargers, each with a twin entry turbine, could each be connected to a group of six cylinders. This would make installation easier, and the frictional losses would be reduced by the shorter pipe lengths. For large marine diesel engines there can be one turbocharger per pair of cylinders. While there are thermodynamic advantages in lagging the turbines and pipework, the ensuing reduction in engine room temperature may be a more important consideration.

The pressure pulses will be reflected back as compression waves and expansion waves. The exact combination of reflected waves will depend on the pipe junctions and turbine entry. The pipe lengths should be such that there are no undesirable interactions in the chosen speed range. For example, the pressure wave from an opening exhaust valve will be partially reflected as a compression wave by the small turbine entry. If the pipe length is very short the reflected wave will increase the pressure advantageously during the initial blow-down period. A slightly longer pipe, and the delayed reflected wave, will increase the pumping during the exhaust stroke – this increases the turbine output at the expense of increased

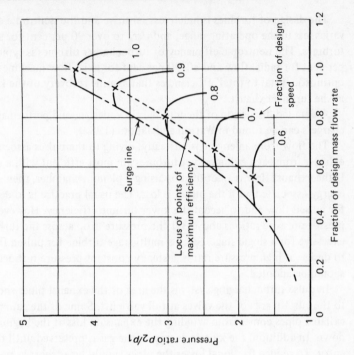

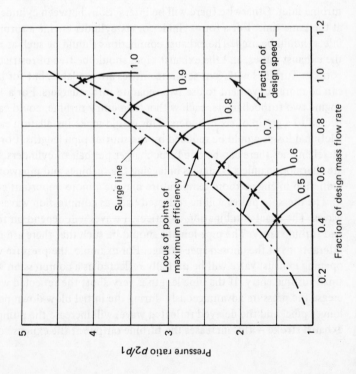

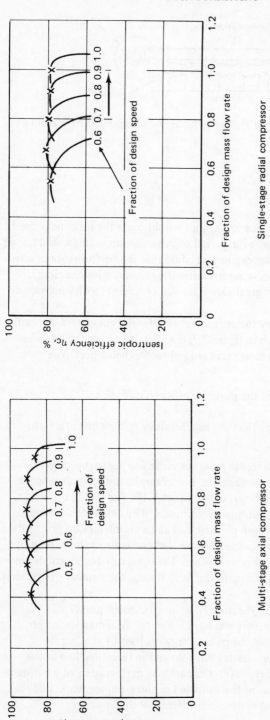

Figure 7.8 Flow characteristics of axial and radial compressors (reproduced with permission from Cohen *et al.* (1972))

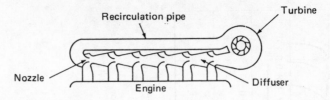

Figure 7.9   Early pulse converter system (with acknowledgement to Watson and Janota (1982))

piston work in the engine. An even longer pipe would cause the reflected wave to return to the exhaust valve during the period of valve overlap — this would impair the performance of a four-stroke engine and could ruin the performance of a two-stroke engine. If the pressure wave returns after the exhaust valve has closed, then it has no effect. Evidently great care is needed on engines with long exhaust pipes and large valve overlaps.

An alternative to multi-entry turbines is the use of pulse converters. An early pulse converter system is shown in figure 7.9; the idea was to use the jet from the nozzle to produce a low-pressure area around each exhaust port. The principal disadvantages are:

(1) insufficient length between the ports for efficient diffusion
(2) high frictional losses
(3) each nozzle has to be larger than the last, resulting in high manufacturing cost.

A more realistic approach is to use pulse converters to connect groups of cylinders that would otherwise be separate. For example, four cylinders could be connected to a single turbocharger entry, figure 7.10. The steadier flow can also lead to an improvement in turbine performance. The design of the pulse converter is a compromise between pressure loss and unwanted pulse propagation. Reducing the throat area increases the pressure loss, but reduces the pulse propagation from one group of cylinders to another. The optimum design will depend on the turbine, the exhaust pipe length, the valve timing, the number of cylinders, the engine speed etc.

Constant-pressure turbocharging (that is, when all exhaust ports enter a chamber at approximately constant pressure) is best for systems with a high pressure ratio. The dissipation of the pulse energy is offset by the improved turbine efficiency. Furthermore, during blow-down the throttling loss at the exhaust valve will be reduced. However, the part load performance of a constant-pressure system is poor because of the increased piston pumping work, and the positive pressure in the exhaust system can interfere with scavenging.

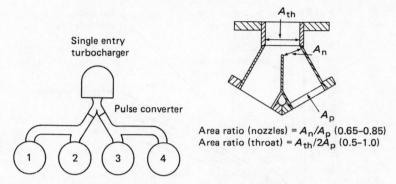

Figure 7.10   Exhaust manifold arrangement (four-cylinder engine) and pulse
               converter details (with acknowledgement to Watson and Janota
               (1982))

7.3 Turbocharging the compression ignition engine

The purpose of turbocharging is to increase the engine output by increasing the
density of the air drawn into the engine. The pressure rise across the compressor
increases the density, but the temperature rise reduces the density. The lower
the isentropic efficiency of the compressor, the greater the temperature rise for
a given pressure ratio.

Substituting for T_{2s} from equation (7.5) into equation (7.3) and rearranging
gives

$$T_2 = T_1 \left[1 + \frac{(p_2/p_1)^{(\gamma-1)/\gamma} - 1}{\eta_c} \right] \tag{7.6}$$

This result is for an ideal gas, and the density ratio can be found by applying
the Gas Law, $\rho = p/RT$. Thus

$$\frac{\rho_2}{\rho_1} = \frac{p_2}{p_1} \left[1 + \frac{(p_2/p_1)^{(\gamma-1)/\gamma} - 1}{\eta_c} \right]^{-1} \tag{7.7}$$

The effect of compressor efficiency on charge density is shown in figure 7.11;
the effect of full cooling (equivalent to isothermal compression) has also been
shown. It can be seen that the temperature rise in the compressor substantially
decreases the density ratio, especially at high pressure ratios. Secondly, the gains
in the density ratio on cooling the compressor delivery can be substantial.
Finally, by ensuring that the compressor operates in an efficient part of the
regime, not only is the work input minimised but the temperature rise is also
minimised. Higher engine inlet temperatures raise the temperature throughout

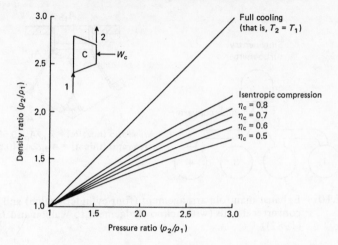

Figure 7.11   Effect of compressor efficiency on air density in the inlet manifold (with acknowledgement to Watson and Janota (1982))

the cycle, and while this reduces ignition delay it increases the thermal loading on the engine.

The advantages of charge cooling lead to the use of inter-coolers. The effectiveness of the inter-cooler can be defined as

$$\epsilon = \frac{\text{actual heat transfer}}{\text{maximum possible heat transfer}}$$

For the cooling medium it is obviously advantageous to use a medium (typically air or water) at ambient temperature (T_1), as opposed to the engine cooling water.

If T_3 is the temperature at exit from the inter-cooler, and the gases are perfect, then

$$\epsilon = \frac{T_2 - T_3}{T_2 - T_1} \tag{7.8}$$

or

$$T_3 = T_2 (1 - \epsilon) + \epsilon T_1$$

In practice it is never possible to obtain heat transfer in a heat exchanger without some pressure drop. For many cases the two are linked linearly by Reynolds' analogy — that is, the heat transfer will be proportional to the pressure drop. In the following simple analysis the pressure drop will be ignored.

Substituting for T_2 from equation (7.6), equation (7.8) becomes

$$T_3 = T_1 \left\{ \left[1 + \frac{(p_2/p_1)^{(\gamma-1)/\gamma} - 1}{\eta_c} \right] (1 - \epsilon) + \epsilon \right\}$$

$$= T_1 \left[1 + (1 - \epsilon) \frac{(p_2/p_1)^{(\gamma-1)/\gamma} - 1}{\eta_c} \right] \tag{7.9}$$

Neglecting the pressure drop in the inter-cooler, equation (7.7) becomes

$$\frac{\rho_3}{\rho_1} = \frac{p_2}{p_1} \left[1 + (1 - \epsilon) \frac{(p_2/p_1)^{(\gamma-1)/\gamma} - 1}{\eta_c} \right]^{-1} \tag{7.10}$$

The effect of charge cooling on the density ratio is shown in figure 7.12 for a typical isentropic compressor efficiency of 70 per cent, and an ambient temperature of 20°C.

Despite the advantages of inter-cooling it is not universally used. The added cost and complexity are not justified for medium output engines, and the provision of a cooling source is troublesome. Gas to gas heat exchangers are bulky and in automotive applications would have to be placed upstream of the radiator. An additional heat exchanger could be used with an intermediate circulating liquid, but with yet more cost and complexity. In both cases energy would be needed to pump the flows. Finally, the added volume of the inter-cooler will influence the transient performance of the engine.

The effect of inter-cooling on engine performance is complex, but two cases will be considered: the same fuelling rate and the same thermal loading. Inter-cooling increases the air flow rate and weakens the air/fuel ratio for a fixed fuelling rate. The temperatures will be reduced throughout the cycle, including the exhaust stage. The turbine output will then be reduced, unless it is rematched, but the compressor pressure ratio will not be significantly reduced. The reduced heat transfer and changes in combustion cause an increase in bmep and a reduction in specific fuel consumption. Watson and Janota (1982) estimate both

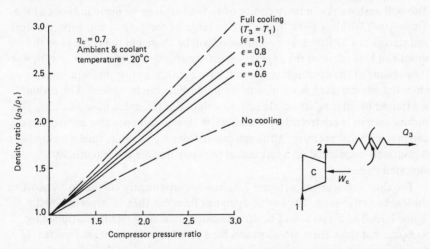

Figure 7.12   Effect of charge cooling on inlet air density (with acknowledgement to Watson and Janota (1982))

changes as 6 per cent for a pressure ratio of 2.25 and inter-cooler effectiveness of 0.7. The gains are greatest at low flow rates where the inter-cooler is most effective.

If the fuelling rate is increased to give the same thermal loading Watson and Janota (1982) estimate a gain in output of 22 per cent. The specific fuel consumption will also be improved since the mechanical losses will not have increased so rapidly as the output.

So far no mention has been made of matching the turbocharger to the engine. Reciprocating engines operate over a wide speed range, and the flow range is further extended in engines with throttle control. In contrast, turbomachinery performance is very dependent on matching the gas flow angles to the blade angles. Consequently, a given flow rate is correct only for a specific rotor speed, and away from the 'design point' the losses increase with increasing incidence angle. Thus turbomachines are not well suited to operating over a wide flow range. However, they do have high design point efficiencies, and are small because of the high speed flows.

The first stage in matching is to estimate the air flow rate. The compressor delivery pressure will be determined by the desired bmep. The air density at entry to the engine (ρ_2) can then be calculated from equation (7.10), and this leads to the air mass flow rate:

$$\dot{m}_a \approx \rho_2 . N^* . V_S . \eta_{\text{vol}}$$

where $\dot{m}_a$ = air mass flow rate (kg/s)
N^* = no. of cycles per second (s^{-1})
V_S = swept volume (m^3)
η_{vol} = volumetric efficiency.

This will enable a preliminary choice of turbocharger to be made in terms of the 'frame size'. Within a given 'frame size', a range of compressor and turbine rotors and stators can be fitted. The compressor will be chosen in the context of the speed and load range of the engine, so that the engine will operate in an efficient flow regime of the compressor, yet still have a sufficient margin from surge. Once the compressor has been matched the turbine can be chosen. The turbine is adjusted by altering its nozzle ring, or volute if it is a radial flow machine. The turbine output is controlled by the effective flow area, hence also controlling the compressor boost pressure. Although calculations are possible, final development is invariably conducted on a test bed in the same manner as for naturally aspirated engines.

The flow characteristics (figure 7.8) can be conveniently combined by plotting contours of efficiency. The engine operating lines can then be superimposed, figure 7.13. The x-axis would be dimensionless mass flow rate if multiplied by $R/A\sqrt{c_p}$, but since these are constants for a given machine they are omitted. If the engine is run at constant speed, but increasing load, then the mass flow rate will increase almost proportionally with the increasing charge density or pressure ratio. This is shown by the nearly vertical straight line in figure 7.13.

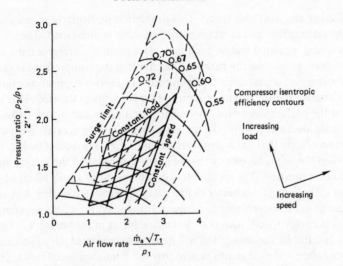

Figure 7.13   Superimposition of engine running lines on compression
               characteristics — constant engine load and speed lines (with
               acknowledgement to Watson and Janota (1982))

If an engine is run at constant load but increasing speed, the volumetric flow
rate of air will also increase. Since the effective flow area of the turbine remains
almost constant, the turbine inlet pressure rises, so increasing the turbine work.
The increased turbine work increases the compressor pressure ratio. This is shown
by the gently rising lines in figure 7.13.

There must be sufficient margin between surge and the nearest operating point
of the engine to allow for two factors. Firstly, the pulsating nature of the flow is
likely to induce surge, and secondly the engine operating conditions may change
from the datum. For example, a blocked air filter or high altitude would reduce
the air flow rate, so moving the operating points closer to surge.

The turbine is tolerant of much wider flow variations than the compressor, and
it is unrealistic to plot mean values for a turbine operating on a pulsed flow. Even
for automotive applications with the widest flow variations, it is usually sufficient
to check only the compressor operation.

The matching of two-stroke engines is simpler, since the flow is controlled by
ports. These behave as orifices in series, and they have a unique pressure drop/
flow characteristic. This gives an almost unique engine operating line, regardless
of speed or load. However, the performance will be influenced by any scavenge
pump.

In automotive applications the wide flow range is made yet more demanding
by the requirement for maximum torque (or bmep) at low speeds. High torque
at low speed reduces the number of gearbox ratios that are needed for starting
and hill climbing. However, if the turbocharger is matched to give high torque at
low speeds, then at high speeds the pressure ratio will be too great, and the

turbocharger may also over-speed. This problem is particularly severe on passenger car engines and an exhaust by-pass valve is often used. The by-pass valve is spring regulated and, at high flow rates when the pressure rises, it allows some exhaust to by-pass the turbine, thus limiting the compressor pressure ratio.

Turbocharging is particularly popular for automotive applications since it enables smaller, lighter and more compact power units to be used. This is essential in cars if the performance of a compression ignition engine is to approach that of a spark ignition engine. In trucks the advantages are even greater. With a lighter engine in a vehicle that has a gross weight limit, the payload can be increased. Also, when the vehicle is empty the weight is reduced, and the vehicle fuel consumption is improved. The specific fuel consumption of a turbocharged compression ignition engine is better than a naturally aspirated engine, but additional gains can be made by retuning the engine. If the maximum torque occurs at an even lower engine speed, then the mechanical losses in the engine will be reduced and the specific fuel consumption will be further improved. However, the gearing will then have to be changed to ensure that the minimum specific fuel consumption occurs at the normal operating point. Ford (1982) claim that turbocharging can reduce the weight of truck engines by 30 per cent, and improve the specific fuel consumption by from 4 to 16 per cent. Figure 7.14 shows a comparison of naturally aspirated and turbocharged truck engines of equivalent power outputs.

In passenger cars a turbocharged compression ignition engine can offer a performance approaching that of a comparably sized spark ignition engine; its torque will be greater but its maximum speed lower. Compression ignition engines can give a better fuel consumption than spark ignition engined vehicles, but this will depend on the driving pattern (Radermacher (1982)) and whether the comparison uses a volumetric or gravimetric basis (see chapter 3, section 3.7).

7.4 Turbocharging the spark ignition engine

Turbocharging the spark ignition engine is more difficult than turbocharging the compression ignition engine. The material from the previous section applies, but in addition spark ignition engines require a wider air flow range (owing to a wider speed range and throttling), a faster response, and more careful control to avoid either pre-ignition or self-ignition (knock). The fuel economy of a spark ignition engine is not necessarily improved by turbocharging. To avoid both knock and self-ignition it is common practice to lower the compression ratio, thus lowering the cycle efficiency. This may or may not be offset by the frictional losses representing a smaller fraction of the engine output.

The turbocharger raises the temperature and pressure at inlet to the spark ignition engine, and consequently pressures and temperatures are raised throughout the ensuing processes. The effect of inlet pressure and temperature on the

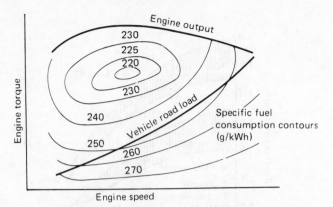

Naturally aspirated 8-cylinder Diesel engine

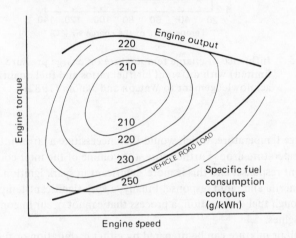

Turbocharged 6-cylinder Diesel engine

Figure 7.14   Comparison of comparably powerful naturally aspirated and turbocharged engines (Ford (1982))

knock-limited operation of an engine running at constant speed, with a constant compression ratio, is shown in figure 7.15. Higher octane fuels and rich mixtures both permit operation with higher boost pressures and temperatures. Retarding the ignition timing will reduce the peak pressures and temperatures to provide further control on knock. Unfortunately there will be a trade-off in power and economy and the exhaust temperature will be higher; this can cause problems with increased heat transfer in the engine and turbocharger. Reducing the compression ratio is the commonest way of inhibiting knock and retarding the ignition is used to ensure knock-free operation under all conditions.

Inter-cooling may appear attractive, but in practice it is very rarely used. Compared to a compression ignition engine, the lower pressure ratios cause a

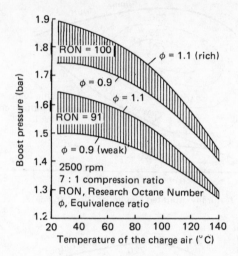

Figure 7.15   Influence of charge temperature on charge pressure (knock-limited) with different air/fuel ratios and fuel qualities (with acknowledgement to Watson and Janota (1982))

lower charge temperature, which would then necessitate a larger inter-cooler for a given temperature drop. Furthermore, the volume of the inter-cooler impairs the transient response, and this is more significant in spark ignition engines with their low inertia and rapid response. Finally, a very significant temperature drop occurs through fuel evaporation, a process that cannot occur in compression ignition engines.

The fuel/air mixture can be prepared by either carburation or fuel injection, either before or after the turbocharger. Fuel injection systems are simplest since they deduce air mass flow rate and will be designed to be insensitive to pressure variations. In engines with carburettors it may appear more attractive to keep the carburettor and inlet manifold from the naturally aspirated engine. However, the carburettor then has to deal with a flow of varying pressure. The carburettor can be rematched by changing the jets, and the float chamber can be pressurised. Unfortunately, it is difficult to obtain the required mixture over the full range of pressures and flow rates. In general it is better to place the carburettor before the compressor for a variety of reasons. The main complication is that the compressor rotor seal needs improvement to prevent dilution of the fuel/air mixture at part load and idling conditions. The most effective solution is to replace the piston ring type seals with a carbon ring lightly loaded against a thrust face. A disadvantage of placing the carburettor before the compressor is that the volume of air and fuel between the carburettor and engine is increased. This can cause fuel hold up when the throttle is opened, and a rich mixture on over-run when the throttle is closed, as discussed in chapter 4, section 4.4.

The advantages of placing the carburettor before the compressor are:

(i) the carburettor operates at ambient pressure
(ii) there is reduced charge temperature
(iii) compressor operation is further from the surge limit
(iv) there is a more homogeneous mixture at entry to the cylinders.

If the carburettor operates at ambient pressure then the fuel pump can be standard and the carburettor can be re-jetted or changed to allow for the increased volumetric flow rate.

The charge temperature will be lower if the carburettor is placed before the compressor. Assuming constant specific heat capacities, and a constant enthalpy of evaporation for the fuel, then the temperature drop across the carburettor (ΔT_{carb}) will be the same regardless of the carburettor position. The temperature rise across the compressor is given by equation (7.6).

$$T_2 = T_1 \left[1 + \frac{(p_2/p_1)^{(\gamma-1)/\gamma} - 1}{\eta_c} \right]$$

The term in square brackets is greater than unity, so that ΔT_{carb} will be magnified if the carburettor is placed before the compressor. In addition, the ratio of the specific heat capacities (γ) will be reduced by the presence of the fuel, so causing a further lowering of the charge temperature. This is illustrated by example 7.2, which also shows that the compressor work will be slightly reduced. The reduced charge temperature is very important since it allows a wider knock-free operation — see figure 7.14.

In spark ignition engines the compressor operates over a wider range of flows, and ensuring that the operation is always away from the surge line can be a greater problem than in compression ignition engines. If the carburettor, and thus the throttle, is placed before the compressor the surge margin is increased at part throttle. Consider a given compressor pressure ratio and mass flow rate and refer back to figure 7.13. The throttle does not change the temperature at inlet to the compressor (T_1), but it reduces the pressure (p_1) and will thus move the operating point to the right of the operating point when the throttle is placed after the compressor and p_1 is not reduced.

By the time a fuel/air mixture passes through the compressor it will be more homogeneous than at entry to the compressor. Furthermore, the flow from the compressor would not be immediately suitable for flow through a carburettor.

The transient response of turbocharged engines is discussed in detail by Watson and Janota (1982). The problems are most severe with spark ignition engines because of their wide speed range and low inertia; the problems are also significant with the more highly turbocharged compression ignition engines. The poor performance under changing speed or load conditions derives from the nature of the energy transfer between the engine and the turbocharger. When the engine accelerates or the load increases, only part of the energy available at the turbine appears as compressor work, the balance is used in accelerating the turbocharger rotor. Additional lags are provided by the volumes in the inlet and exhaust systems between the engine and turbocharger; these volumes should be minimised for good transient response. Furthermore, the inlet volume should be minimised in spark ignition engines to limit the effect of fuel hold-up on the fuel-wetted surfaces. Turbocharger lag cannot be eliminated without some additional energy input, but the effect can be minimised. One approach is to under-size the turbocharger, since the rotor inertia increases with $(length)^4$, while the flow area increases with $(length)^2$. Then to prevent undue back-pressure in the exhaust, an exhaust by-pass valve can be fitted. An alternative approach is to replace a single turbocharger by two smaller units.

The same matching procedure is used for spark ignition engines and compression ignition engines. However, the wider speed and flow range of the spark ignition engine necessitates greater compromises in the matching of turbo-machinery to a reciprocating engine. If the turbocharger is matched for the maximum flow then the performance at low flows will be very poor, and the large turbocharger size will give a poor transient response. When a smaller turbocharger is fitted, the efficiency at low flow rates will be greater and the boost pressure will be higher throughout the range; the lower inertia will also reduce turbocharger lag. However, at higher flow rates the boost pressure would become excessive unless modified; two approaches are shown in figure 7.16.

The compressor pressure can be directly controlled by a relief valve, to keep the boost pressure below the knock-limited value. The flow from the relief valve does not represent a complete loss of work since the turbine work derives from energy that would otherwise be dissipated during the exhaust blow-down. The blow-off flow can be used to cool the turbine and exhaust systems. If the carburettor is placed before the compressor, then the blow-off flow has to be returned to the compressor inlet, which results in yet higher charge temperatures.

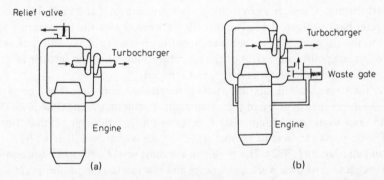

Figure 7.16   (a) Compressor pressure-relief valve control system. (b) Boost pressure-sensitive waste control system (with acknowledgement to Watson and Janota (1982))

The exhaust waste gate system, figure 7.16b, is more attractive since it also permits a smaller turbine to be used, because it no longer has to be sized for the maximum flow. Turbocharger lag is reduced by the low inertia, and the control system ensures that the waste gate closes during acceleration. The main difficulty is in designing a cheap reliable system that will operate at the high temperatures. Variable area turbines, compressor restrictors and turbine outlet restrictors can also be used to control the boost pressure. A variable area turbine is not sufficiently better than a waste gate, so its use is not justified; restrictors of any form are an unsatisfactory solution.

Performance figures vary, but typically a maximum boost pressure of 1.5 bar would raise the maximum torque by 30 per cent and maximum power by up to 60 per cent. Figure 7.17 shows the comparative specific fuel consumption of a turbocharged and naturally aspirated spark ignition engine. The turbocharged engine has improved fuel consumption at low outputs, but an inferior consumption at higher outputs. The effect on vehicle consumption would depend on the particular driving pattern.

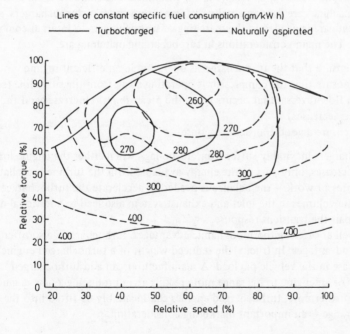

Figure 7.17   Comparative specific fuel consumption of a turbocharged and naturally aspirated petrol engine scaled for the same maximum torque (with acknowledgement to Watson and Janota (1982))

7.5 Conclusions

Turbocharging is a very important means of increasing the output of internal combustion engines. Significant increases in output are obtained, yet the turbocharger system leads to only small increases in the engine weight and volume.

The fuel economy of compression ignition engines is usually improved by turbocharging, since the mechanical losses do not increase in direct proportion to the gains in power output. The same is not necessarily true of spark ignition engines, since turbocharging invariably necessitates a reduction in compression ratio to avoid knock (self-ignition of the fuel/air mixture). The reduction in compression ratio reduces the indicated efficiency and this usually negates any improvement in the mechanical efficiency.

The relatively low flow rate in turbochargers leads to the use of radial flow compressors and turbines. In general, axial flow machines are more efficient, but only for high flow rates. Only in the largest turbochargers (such as those for marine applications) are axial flow turbines used. Turbochargers are unlike positive displacement machines, since they rely on dynamic flow effects; this implies high velocity flows, and consequently the rotational speeds are an order of magnitude greater than reciprocating machines. The characteristics of reciprocating machines are fundamentally different from those of turbochargers, and thus great care is needed in the matching of turbochargers to internal combustion engines. The main considerations in turbocharging matching are:

(i)   to ensure that the turbocharger is operating in an efficient regime
(ii) to ensure that the compressor is operating away from the surge line (surge is a flow reversal that occurs when the pressure ratio increases and the flow rate decreases)
(iii) to ensure a good transient response.

Turbochargers inevitably suffer from 'turbo-lag'; when either the engine load or speed increases, only part of the energy available from the turbine is available as compressor work — the balance is needed to accelerate the turbocharger rotor. The finite volumes in the inlet and exhaust system also lead to additional delays that impair the transient response.

As well as offering thermodynamic advantages, turbochargers also offer commercial advantages. In trucks, the reduced weight of a turbocharged engine gives an increase in the vehicle payload. A manufacturer can add turbocharged versions of an engine to his range more readily than producing a new engine series. Furthermore, turbocharged engines can invariably be fitted into the same vehicle range — an important marketing consideration.

7.6 Examples

Example 7.1

A diesel engine is fitted with a turbocharger, which comprises a radial compressor driven by a radial exhaust gas turbine. The air is drawn into the compressor at a pressure of 0.95 bar and at a temperature of 15°C, and is delivered to the engine at a pressure of 2.0 bar. The engine is operating on a gravimetric air/fuel ratio of 18:1, and the exhaust leaves the engine at a temperature of 600°C and at a pressure of 1.8 bar; the turbine exhausts at 1.05 bar. The isentropic efficiencies of the compressor and turbine are 70 per cent and 80 per cent, respectively. Using the values;

$$c_{p_{air}} = 1.01 \text{ kJ/kg K}, \quad \gamma_{air} = 1.4$$

and

$$c_{p_{ex}} = 1.15 \text{ kJ/kg K}, \quad \gamma_{ex} = 1.33$$

calculate   (i) the temperature of the air leaving the compressor

         (ii) the temperature of the gases leaving the turbine

       (iii) the mechanical power loss in the turbocharger expressed as a percentage of the power generated in the turbine.

Referring to figure 7.18 (a new version of figure 7.6), the real and ideal temperatures can be evaluated along with the work expressions.

(i) If the compression were isentropic, $T_{2s} = T_1 \left(\dfrac{p_2}{p_1}\right)^{(\gamma-1)/\gamma}$

$$T_{2s} = 288 \left(\frac{2.0}{0.95}\right)^{(1.4-1)/1.4} = 356 \text{ K, or } 83°C$$

From the definition of compressor isentropic efficiency, $\eta_c = \dfrac{T_{2s} - T_1}{T_2 - T_1}$

$$T_2 = \frac{T_{2s} - T_1}{\eta_c} + T_1 = \frac{83 - 15}{0.7} + 15 = 113°C$$

(ii) If the turbine were isentropic, $T_{4s} = T_3 \left(\dfrac{p_4}{p_3}\right)^{(\gamma-1)/\gamma}$

$$T_{4s} = 873 \left(\frac{1.05}{1.8}\right)^{(1.33-1)/1.33} = 762.9 \text{ K or } 490°C$$

From the definition of turbine isentropic efficiency, $\eta_t = \dfrac{T_3 - T_4}{T_3 - T_{4s}}$

$$T_4 = T_3 - \eta_t (T_3 - T_{4s}) = 600 - 0.8 (600 - 490) = 512°C$$

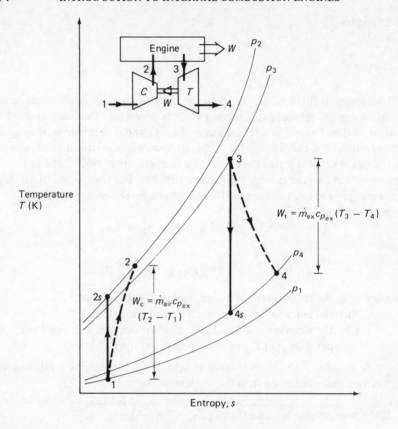

Figure 7.18   Temperature/entropy diagram for a turbocharger

(iii) Compressor power $\dot{W}_c = \dot{m}_{air} \, c_{p_{air}} \, (T_2 - T_1)$

$$= \dot{m}_{air} \, 1.01 \, (113 - 15) \text{ kW}$$

$$= \dot{m}_{air} \, 98.98 \text{ kW}$$

from the air/fuel ratio

$$\dot{m}_{ex} = \dot{m}_{air} \left(1 + \frac{1}{18}\right)$$

and turbine power

$$\dot{W}_t = \dot{m}_{ex} \, c_{p_{ex}} \, (T_3 - T_4)$$

$$= \dot{m}_{air} \, 1.056 \times 1.15 \, (600 - 512)$$

$$= \dot{m}_{air} \, 106.82 \text{ kW}$$

Thus, the mechanical power loss as a percentage of the power generated in the turbine is

$$\frac{106.82 - 98.98}{106.82} \times 100 = 7.34 \text{ per cent}$$

This result is in broad agreement with figure 7.7, which is for a slightly different pressure ratio and constant gas flow rates and properties.

Example 7.2

Compare the cooling effect of fuel evaporation on charge temperature in a turbocharged spark ignition engine for the following two cases:

(a) the carburettor placed before the compressor
(b) the carburettor placed after the compressor.

The specific heat capacity of the air and the latent heat of evaporation of the fuel are both constant. For the air/fuel ratio of 12.5:1 the evaporation of the fuel causes a 25 K drop in mixture temperature. The compressor efficiency is 70 per cent for the pressure ratio of 1.5, and the ambient air is at 15°C. Assume the following property values:

for air $c_p = 1.01$ kJ/kg K, $\gamma = 1.4$
for air/fuel mixture $c_p = 1.05$ kJ/kg K, $\gamma = 1.34$

Finally, compare the compressor work in both cases.
Both arrangements are shown in figure 7.19.

(a) $T_1 = 15°C = 288$ K
$T_2 = T_1 - 25 = 263$ K

If the compressor were isentropic, $T_{3s} = 263 (1.5)^{(1.34-1)/1.34} = 291.5$ K

From the definition of compressor isentropic efficiency

$$T_3 = \frac{T_{3s} - T_2}{\eta_c} + T_2 = \frac{291.5 - 263}{0.7} + 263 = 303.7 \text{ K}$$

(b) $T_4 = 288$ K

For isentropic compression $T_{5s} = 288 (1.5)^{(1.4-1)/1.4} = 323.4$ K

From the definition of compressor isentropic efficiency

$$T_5 = \frac{T_{5s} - T_4}{\eta_c} + T_4 = \frac{323.4 - 288}{0.7} + 288 = 338.5 \text{ K}$$

$$T_6 = T_5 - 25 = 338.5 - 25 = 313.5 \text{ K}$$

Since $T_6 > T_3$, it is advantageous to place the carburettor before the compressor.

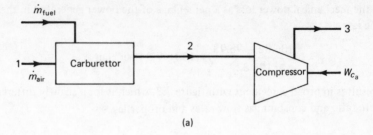

(a)

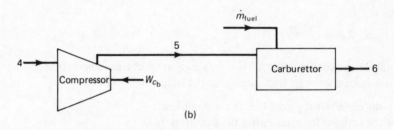

(b)

Figure 7.19   Possible arrangement for the carburettor and compressor in a
spark ignition engine. (a) Carburettor placed before the compressor; (b) carburettor placed after the compressor

Comparing the compressor power for the two cases:

$$(W_c)_a = \dot{m}_{mix} \, c_{p_{mix}} \, (T_3 - T_2)$$

$$= 1.08 \, \dot{m}_{air} \, 1.05 \, (303.7 - 263)$$

$$= 46.15 \, \dot{m}_{air} \, kW$$

$$(W_c)_b = \dot{m}_{air} \, c_{p_{air}} \, (T_5 - T_4)$$

$$= \dot{m}_{air} \, 1.01 \, (338.5 - 288)$$

$$= 51.01 \, \dot{m}_{air} \, kW$$

Thus placing the carburettor before the compressor offers a further advantage in
reduced compressor work.

The assumptions in this example are somewhat idealised. When the carburettor
is placed before the compressor, the fuel will not be completely evaporated
before entering the compressor, and evaporation will continue during the compression process. However, less fuel is likely to enter the cylinders in droplet
form if the carburettor is placed before the compressor rather than after.

7.7 Problems

7.1 A spark ignition engine is fitted with a turbocharger that comprises a radial flow compressor driven by a radial flow exhaust gas turbine. The gravimetric air/fuel ratio is 12:1, with the fuel being injected between the compressor and the engine. The air is drawn into the compressor at a pressure of 1 bar and at a temperature of 15°C. The compressor delivery pressure is 1.4 bar. The exhaust gases from the engine enter the turbine at a pressure of 1.3 bar and a temperature of 710°C; the gases leave the turbine at a pressure of 1.1 bar. The isentropic efficiencies of the compressor and turbine are 75 per cent and 85 per cent, respectively.

Treating the exhaust gases as a perfect gas with the same properties as air, calculate:

  (i) the temperature of the gases leaving the compressor and turbine
 (ii) the mechanical efficiency of the turbocharger.

7.2 Why is it more difficult to turbocharge spark ignition engines than compression ignition engines? Under what circumstances might a supercharger be more appropriate?

7.3 Why do compression ignition engines have greater potential than spark ignition engines for improvements in power output and fuel economy as a result of turbocharging? When is it most appropriate to specify an intercooler?

7.4 Derive an expression that relates compressor delivery pressure (p_2) to turbine inlet pressure (p_3) for a turbocharger with a mechanical efficiency η_{mech}, and compressor and turbine isentropic efficiencies η_c and η_t, respectively. The compressor inlet conditions are p_1, T_1, the turbine inlet temperature is T_3 and the outlet pressure is p_4. The air fuel ratio (AFR) and the differences between the properties of air (suffix a) and exhaust (suffix e) must all be considered.

7.5 Why do turbochargers most commonly use radial flow compressors and turbines with non-constant pressure supply to the turbine?

7.6 Why does turbocharging a compression ignition engine normally lead to an improvement in fuel economy, while turbocharging a spark ignition engine usually leads to decreased fuel economy?

8 Mechanical Design Considerations

8.1 Introduction

Once the type and size of engine have been determined, the number and disposition of the cylinders have to be decided. Very often the decision will be influenced by marketing and packaging considerations, as well as whether or not the engine needs to be manufactured with existing machinery.

The engine block and cylinder head are invariably cast, the main exception being the fabrications used for some large marine diesel engines. The material is usually cast iron or an aluminium alloy. Cast iron is widely used since it is cheap and easy to cast; once the quenched outer surfaces have been penetrated it is also easy to machine. Aluminium alloys are more expensive but lighter, and are thus likely to gain in importance as designers seek to reduce vehicle weight.

Pistons are invariably made from an aluminium alloy, but in higher-output compression ignition engines the piston crown needs to be protected by either a cast iron or ceramic top. The piston rings are cast iron, sometimes with a chromium-plated finish. The valves are made from one or more alloy steels to ensure adequate life under their extreme operating conditions.

Engine bearings are invariably of the journal type with a forced lubrication system. To economise on the expensive bearing alloys, thin-wall or shell-type bearings are used; these have a thin layer of bearing metal on a strip steel backing. These bearings can easily be produced in two halves, making assembly and replacement of all the crankshaft bearings (main and big-end) very simple. For the more lightly loaded bearings the need for separate bearing materials can be eliminated by careful design. The use of roller or ball bearings in crankshafts is limited because of the ensuing need for a built-up crankshaft; the only common application is in some motorcycle engines.

The role of the lubricant is not just confined to lubrication. The oil also acts as a coolant (especially in some air-cooled engines), as well as neutralising the effects of the corrosive combustion products.

Only an outline of the main mechanical design considerations can be given here. Further information can be found in the SAE publications, and books such as those detailed in Baker (1979), Newton *et al.* (1983) and Taylor (1968).

8.2 The disposition and number of the cylinders

The main constraints influencing the number and disposition of the cylinders are:

(1) the number of cylinders needed to produce a steady output
(2) the minimum swept volume for efficient combustion (say 400 cm^3)
(3) the number and disposition of cylinders for satisfactory balancing.

The most common engine types are: the straight or in-line, the 'V' (with various included angles), and the horizontally opposed – see figure 8.1.

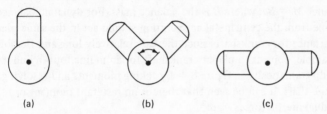

(a)   (b)   (c)

Figure 8.1 Common engine arragements. (a) In-line; (b) 'V'; (c) horizontally opposed

'V' engines form a very compact power unit; a more compact arrangement is the 'H' configuration (in effect two horizontally opposed engines with the crankshafts geared together), but this is an expensive and complicated arrangement that has had limited use. Whatever the arrangement, it is unusual to have more than six or eight cylinders in a row because torsional vibrations in the crankshaft then become much more troublesome. In multi-cylinder engine configurations other than the in-line format, it is advantageous if a crankpin can be used for a connecting-rod to each bank of cylinders. This makes the crankshaft simpler, reduces the number of main bearings, and facilitates a short crankshaft that will be less prone to torsional vibrations. None the less, the final decision on the engine configuration will also be influenced by marketing, packaging and manufacturing constraints.

In deciding on an engine layout there are the two inter-related subjects; the engine balance and the firing interval between cylinders. The following discussions will relate to four-stroke engines, since these only have a single firing stroke in each cylinder once every two revolutions. An increase in the number of cylinders leads to smaller firing intervals and smoother running, but above six cylinders the improvements are less noticeable. Normally the crankshaft is arranged to give equal firing intervals, but this is not always the case. Sometimes a compromise is made for better balance or simplicity of construction; for example, consider a twin cylinder horizontally opposed four-stroke engine with a single throw crankshaft – the engine is reasonably balanced but the firing intervals are 180°, 540°, 180°, 540° etc.

The subject of engine balance is treated very thoroughly by Taylor (1968), with the results tabulated for the more common engine arrangements.

When calculating the engine balance, the connecting-rod is treated as two masses concentrated at the centre of the big-end and the centre of the little-end – see figure 8.2. For equivalence

$$m_1 = m_1 + m_2$$

and $$m_1 r_1 = m_2 r_2$$

The mass m_2 can be considered as part of the mass of the piston assembly (piston, rings, gudgeon pin etc.) and be denoted by m_r, the reciprocating mass. The crankshaft is assumed to be in static and dynamic balance, figure 8.3. For static balance $Mr = Ba$, where B is the balance mass. For dynamic balance the inertia force from the centripetal acceleration should act in the same plane; this is of importance for crankshafts since they are relatively long and flexible. As a simple example, consider a planar crankshaft for an in-line four-cylinder engine, as shown diagramatically in figure 8.4. By taking moments and resolving at any point on the shaft, it can be seen that there is no resultant moment or force from the individual inertia forces $mr\omega^2$.

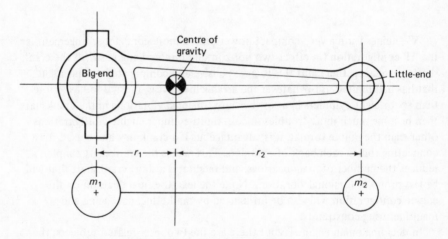

Figure 8.2 Connecting-rod and its equivalent

The treatment of the reciprocating mass is more involved. If the connecting-rod were infinitely long the reciprocating mass would follow simple harmonic motion, producing a primary out-of-balance force. However, the finite length of the connecting-rod introduces higher harmonic forces. A full analysis is given by Taylor (1968), who shows that only the primary and secondary harmonic terms need be considered:

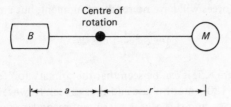

Figure 8.3 Balancing arrangements

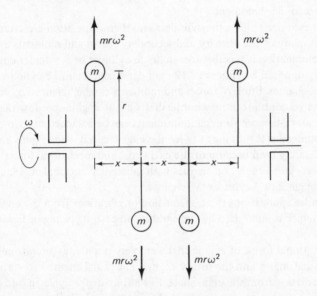

Figure 8.4 Diagrammatic representation of a four-cylinder in-line engine crankshaft

$$F_r \approx m_r\omega^2 r \left(\cos\theta + \frac{r}{l}\cos 2\theta\right)$$

where F_r = axial force due to the reciprocating mass
        m_r = equivalent reciprocating mass
        ω = angular velocity, $d\theta/dt$
        r = crankshaft throw
        l = connecting-rod length
  $\cos\theta$ = primary term
 $\cos 2\theta$ = secondary term

In other words there is a primary force varying in amplitude with crankshaft rotation and a secondary force varying at twice the crankshaft speed; these forces act along the cylinder axis. Referring back to figure 8.4, it can be seen that the primary forces will have no resultant force or moment.

The secondary forces will have no resultant moment, but a resultant force of

$$4m_r \omega^2 \ \frac{r^2}{l} \ \cos 2\theta$$

By referring to figure 8.5, it can be seen that the primary forces for this 4-cylinder engine are 180° out of phase and thus cancel. However, the secondary forces will be in phase, and this causes a resultant secondary force on the bearings. Since the resultant secondary forces have the same magnitude and direction there is no secondary moment.

In multi-cylinder engines the cylinders and their disposition are arranged to eliminate as many of the primary and secondary forces and moments as possible. Complete elimination is possible for: in-line 6-cylinder or 8-cylinder engines, horizontally opposed 8-cylinder or 12-cylinder engines, and 12-cylinder or 16-cylinder 'V' engines. Primary forces and moments can be balanced by masses running on two contra-rotating countershafts at the engine speed, while secondary out-of-balance forces and moments can be balanced by two contra-rotating countershafts running at twice the engine speed — see figure 8.6. Such systems are rarely used because of the extra cost and mechanical losses involved, but examples can be found on engines with inherently poor balance such as in-line 3-cylinder engines or 4-cylinder 'V' engines.

In vehicular applications the transmission of vibrations from the engine to the vehicle structure is minimised by the careful choice and placing of flexible mounts.

Two additional forms of engine that were very popular as aircraft engines were the radial engine and the rotary engine. The radial engine had stationary cylinders radiating from the crankshaft. For a four-stroke cycle an odd number of cylinders is used in order to give equal firing intervals. If there are more than three rows of cylinders there are no unbalanced primary or secondary forces or moments. In the rotary engine the cylinders were again radiating from the crankshaft, but the cylinders rotated about a stationary crankshaft.

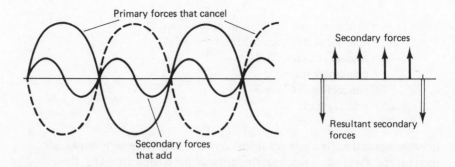

Figure 8.5   Secondary forces for the crankshaft shown in figure 8.4

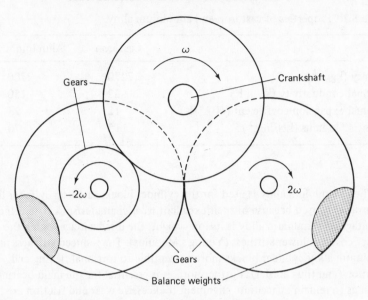

Figure 8.6 Countershafts for balancing secondary forces

8.3 Cylinder block and head materials

Originally the cylinder head and block were often an integral iron casting. By eliminating the cylinder head gasket, problems with distortion, thermal conduction between the block and head, and gasket failure were avoided. However, manufacture and maintenance were more difficult. The most widely used materials are currently cast iron and aluminium alloys. Typical properties are shown in table 8.1.

There are several advantages associated with using an aluminium alloy for the cylinder head. Aluminium alloys have the advantage of lightness in weight and ease of production to close tolerances by casting — very important considerations for the combustion chambers. The high thermal conductivity also allows higher compression ratios to be used, because of the reduced problems associated with hot spots. The main disadvantages are the greater material costs, the greater susceptibility to damage (chemical and mechanical), and the need for valve seat inserts and valve guides (see figure 6.2). Furthermore, the mechanical properties of aluminium alloys are poorer than cast iron. The greater coefficient of thermal expansion and the lower Young's Modulus make the alloy cylinder head more susceptible to distortion. None the less, aluminium alloy is increasingly being used for cylinder heads.

Table 8.1 Properties of cast iron and aluminium alloy

	Cast iron	Aluminium alloy
Density (kg/m^3)	7270	2700
Thermal conductivity (W/m K)	52	150
Thermal expansion coefficient (10^{-6}/K)	12	23
Young's Modulus (kN/mm^2)	115	70

When aluminium alloy is used for the cylinder block, cast iron cylinder liners are invariably used because of their excellent wear characteristics. The principal advantage of aluminium alloy is its low weight, the disadvantages being the greater cost and lower stiffness (Young's Modulus). The reduced stiffness makes aluminium alloy cylinder blocks more susceptible to torsional flexing and vibration (and thus noisier). Furthermore, it is essential for the main bearing housings to remain in accurate alignment if excessive wear and friction are to be eliminated. These problems are overcome by careful design, with ribs and flanges increasing the stiffness.

In order to facilitate design, much use is now made of finite element methods. These enable the design to be optimised by carrying out stress analysis and vibration analysis with different arrangements and thicknesses of ribs and flanges. In addition, the finite element method can be applied to heat transfer problems, and the thermal stresses can be deduced to complete the model.

An interesting example of an aluminium cylinder block is the Chevrolet Vega 2.3 litre engine. The open-deck design reduces the torsional stiffness of the block but enables the block to be diecast, thus greatly easing manufacture. The aluminium alloy contains 16–18 per cent silicon, 4–5 per cent copper and 0.45–0.65 per cent magnesium. Cast iron cylinder liners are not used; instead the cylinder bore is treated to form a wear-resistant and oil-retaining surface by electrochemical etching to expose the hard silicon particles. To provide a compatible bearing surface the piston skirts are electroplated successively with zinc, copper, iron and tin. The zinc bonds well to the piston alloy, and the copper protects the zinc; the iron provides the bearing material, while the tin protects the iron and facilitates the running-in.

In larger (non-automotive) engines, steel liners are often used because of their greater strength compared to that of cast iron. To provide an inert, oil-retaining, wear-resistant surface a carefully etched chromium-plated finish is often used.

The majority of engines are water-cooled, and the coolant passages are formed by sand cores during casting. Water is a very effective heat transfer medium, not least because, if there are areas of high heat flux, nucleate boiling can occur locally, thereby removing large quantities of energy without an excessive temperature rise. Because of the many different materials in the cooling system it is always advisable to use a corrosion inhibitor, such as those that are added to ethylene glycol in antifreeze mixtures.

A typical cooling-water arrangement is shown in figure 8.7. This is for the Ford direct injection compression ignition engine that is discussed in chapter 1, section 1.3, and shown in figure 1.10. At the front of the engine is a water pump which enhances the natural convection flow in the cooling system. The pump is driven by a V-belt from the crankshaft pulley. The outflow from the pump is divided into two flows which enter opposite ends of the cylinder block. On four-cylinder engines the flow would not normally be divided in this way. The flow to the far end of the engine first passes through the oil cooler. Once the water enters the cylinder block it passes around the cylinders, and rises up into the cylinder head. While the engine is reaching its working temperature the flow passes straight to the pump inlet since the thermostats are closed.

When the engine approaches its working temperature first one thermostat opens and then, at a slightly higher temperature (say 5 K), the second thermostat opens. Once the second thermostat has opened about 50 per cent of the flow passes through the radiator. The use of two thermostats prevents a sudden surge of cold water from the radiator, and also provides a safety margin should

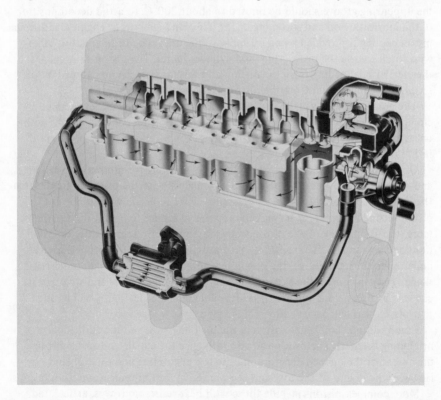

Figure 8.7 Cooling-water system; see also figures 1.10 and 8.16 (courtesy of Ford)

one thermostat fail. The optimum water flow pattern is often obtained from experiments with clear plastic models.

Air-cooling is also used for reasons of lightness, compactness and simplicity. However, the higher operating temperatures require a more expensive construction, and there is more noise from the combustion, pistons and fan.

8.4 The piston and rings

The design and production of pistons and rings is a complicated job, which is invariably carried out by specialist manufacturers; a piston assembly is shown in figure 8.8.

Pistons are invariably made from aluminium alloy, a typical composition being 10–12 per cent silicon to give a relatively low coefficient of thermal expansion of 19.5×10^{-6} K^{-1}. The low density reduces the reciprocating mass, and the good thermal conductivity avoids hot spots. The temperature of the piston at the upper ring groove should be limited to about 200°C, to avoid decomposition of the lubricating oil and softening of the alloy. In high-output engines, additional piston cooling is provided by an oil spray to the underside of the piston; otherwise cooling is via the piston rings and cylinder barrel. The piston skirt carries the inertial side loading from the piston, and this loading can be reduced by offsetting the gudgeon pin from the piston diameter.

The gudgeon pin is usually of hollow, case hardened steel, either retained by circlips or of an accurate diametral fit. The centre hole reduces the weight without significantly reducing the strength. The piston is reinforced by bosses in the region of the gudgeon pin.

One of the key problems in piston design is allowing for thermal expansion and distortion. The thermal coefficient of expansion for the piston is greater than that of the bore, so that sufficient clearances have to be allowed to prevent the piston seizing when it is at its maximum possible service temperature. Furthermore, the asymmetry of the piston leads to non-uniform temperature distributions and asymmetrical expansions. To ensure minimal but uniform clearances under operating conditions, the piston is accurately machined to a non-circular shape. To help control the expansion, carefully machined slots and steel inserts can also be used. None the less, it is inevitable that the clearances will be such that piston slap will occur with a cold engine.

Combustion chambers are often in the piston crown, and the additional machining is trivial. The piston can also influence the engine emissions through the extent of the quench areas around the top piston ring and the top land. However the extent of the top land is governed by piston-temperature limitations.

More complex pistons include those with heat-resistant crowns, articulated skirts, and raised pads on the skirt to reduce the frictional losses.

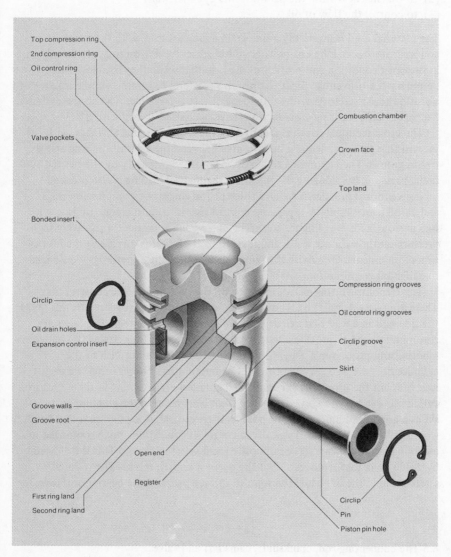

Figure 8.8   Piston assembly (with acknowledgement to GKN Engine Parts
             Division)

The three main roles of the piston rings are:

(1) to seal the combustion chamber
(2) to transfer heat from the piston to the cylinder walls
(3) to control the flow of oil.

The material used is invariably a fine grain alloy cast iron, with the excellent heat and wear resistance inherent in its graphitic structure. Piston rings are usually cast in the open condition and profile-finished, so that when they are closed their periphery is a true circle. Since the piston rings tend to rotate a simple square-cut slot is quite satisfactory, with no tendency to wear a vertical ridge in the cylinder. Numerous different ring cross-sections have been used — see figure 8.9.

The cross-sectional depth is dictated by the required radial stiffness, with the proviso that there is adequate bearing area between the sides of the ring and the piston groove. The piston ring thickness is governed primarily by the desired radial pressure; by reducing the thickness the inertial loading is also reduced.

Conventional practice is to have three piston rings: two compression rings, and an oil control ring. A typical oil control ring is of slotted construction with two narrow lands — see figure 8.10. The narrow lands produce a relatively high pressure on the cylinder walls, and this removes oil that is surplus to the lubrication requirements. Otherwise the pumping action of the upper rings would lead to a high oil consumption.

The Hepolite SE ring shown in figure 8.10 is of a three-piece construction. There are two side rails and a box section with a cantilever torsion bar; this loads the side rails in a way that can accommodate uneven bore wear. The lands are chromium-plated to provide a hard wearing surface. For long engine life between reboring and piston replacement, the material selection and finish of the cylinder bore are also very important. Chromium-plated bores provide a very long life, but the process is expensive and entails a long running-in time. Cast iron cylinder bores are also very satisfactory, not least because the graphite particles act as a good solid lubricant. The bore finish is critical, since too smooth a finish would fail to hold any oil. Typically, a coarse silicon carbide hone is plunged in and out to produce two sets of opposite hand spiral markings. This is followed by a fine hone that removes all but the deepest scratch marks. The residual scratches hold the oil, while the smooth surface provides the bearing surface.

8.5 The connecting-rod, crankshaft, camshaft and valves

Connecting-rods are invariably steel stampings, with an 'H' cross-section centre section to provide high bending strength. Titanium, aluminium alloys and cast irons have all been used for particular applications, with the manufacture being by forging and machining. The big-end bearing is invariably split for ease of

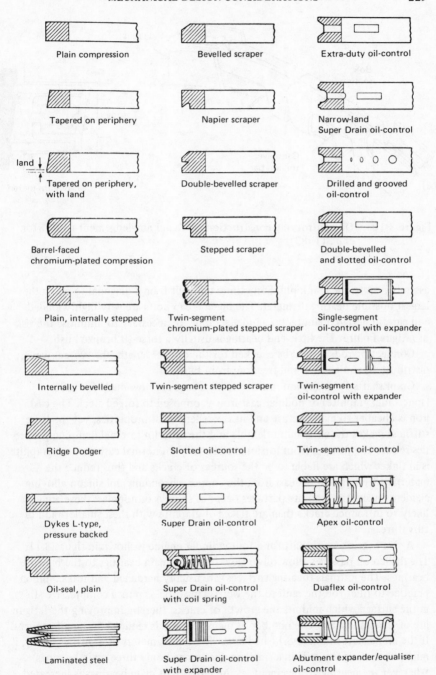

land ↓

Plain compression

Tapered on periphery

Tapered on periphery, with land

Barrel-faced chromium-plated compression

Plain, internally stepped

Internally bevelled

Ridge Dodger

Dykes L-type, pressure backed

Oil-seal, plain

Laminated steel

Bevelled scraper

Napier scraper

Double-bevelled scraper

Stepped scraper

Twin-segment chromium-plated stepped scraper

Twin-segment stepped scraper

Slotted oil-control

Super Drain oil-control

Super Drain oil-control with coil spring

Super Drain oil-control with expander

Extra-duty oil-control

Narrow-land Super Drain oil-control

Drilled and grooved oil-control

Double-bevelled and slotted oil-control

Single-segment oil-control with expander

Twin-segment oil-control with expander

Twin-segment oil-control

Apex oil-control

Duaflex oil-control

Abutment expander/equaliser oil-control

Figure 8.9   Different types of piston ring (with acknowledgement to Baker (1979); and the source of data: AE Group)

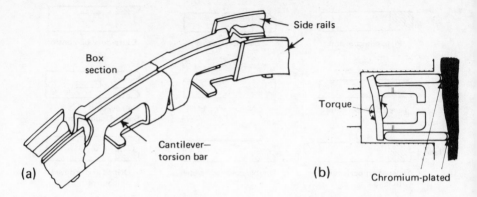

Figure 8.10   Oil control ring construction (with acknowledgement to Newton *et al.* (1983))

assembly on to the crank pin. Sometimes the split is on a diagonal to allow the largest possible bearing diameter. The big-end cap bolts are very highly loaded and careful design, manufacture and assembly are necessary to minimise the risk of fatigue failure. The little-end bearing is usually a force-fit bronze bush.

Connecting-rods should be checked for the correct length, the correct weight distribution, straightness and freedom from twist.

Crankshafts for many automotive applications are now made from SG (spheroidal graphite) or nodular cast iron as opposed to forged steel. The cast iron is cheaper to manufacture and has excellent wear properties, yet the lower stiffness makes the shaft more flexible and the superior internal damping properties reduce the dangers from torsional vibrations. In normal cast iron the graphite is in flakes which are liable to be the sources of cracks and thus reduce the material's strength. In SG cast iron, the copper, chromium and silicon alloying elements make the graphite particles occur as spheres or nodules; these are less likely to introduce cracks than are flakes of graphite with their smaller radii of curvature.

A five-bearing crankshaft for a four-cylinder engine is shown in figure 8.11. The drilled oil passages allow oil to flow from the main bearings to the big-end bearings. The journals (bearing surfaces) are usually hardened and it is common practice to fillet-roll the radii to the webs. This process puts a compressive stress in the surface which inhibits the growth of cracks, thereby improving the fatigue life of the crankshaft. The number of main bearings is reduced in some instances. If the crankshaft in figure 8.11 had larger journal diameters or a smaller throw, it might be sufficiently stiff in a small engine to need only three main bearings. Whatever the bearing arrangement, as the number of main bearings is increased it becomes increasingly important for the journals and main bearings to be accurately in-line.

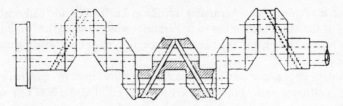

Figure 8.11   Five-bearing crankshaft for a four-cylinder engine (with acknowledgement to Newton *et al.* (1983))

To reduce torsional vibration a damper can be mounted at the front end of
the crankshaft. A typical vibration damper is shown in figure 8.12, where a V-
belt drive has also been incorporated. An annulus is bonded by rubber to the
hub, and the inertia of the annulus and the properties of the rubber insert are
chosen for the particular application. The torsional energy is dissipated as heat
by the hysteresis losses in the rubber. The annulus also acts as a 'vibration
absorber' by changing the crankshaft vibration characteristics.

Camshafts are typically made from hardened steel, hardened alloy cast iron,
nodular cast iron or chilled cast iron. Chill casting is when suitably shaped iron
'chills' are inserted into the mould to cause rapid cooling of certain parts. The
rapid cooling prevents some of the iron carbide dissociating, and thus forms a
very hard surface. A variety of surface hardening techniques are used, including

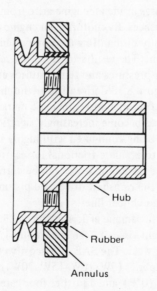

Hub

Rubber

Annulus

Figure 8.12   Torsional vibration damper (with acknowledgement to Newton
et al. (1983))

induction hardening, flame hardening, nitriding, Tufftriding and carburising. The material of the cam follower has to be carefully selected since the components are very highly loaded, and the risk of surface pick-up or cold welding must be minimised.

The inlet valves and in particular the exhaust valves have to operate under arduous conditions with temperatures rising above 500°C and 800°C, respectively. To economise on the exhaust valve materials a composite construction can be used; a Nimonic head with a stellite facing may be friction-welded to a cheaper stem. This also allows a material with a low coefficient of thermal expansion to be used for the stem. The valve guide not only guides the valve, but also helps to conduct heat from the valve to the cylinder head. In cast iron cylinder heads the guide is often an integral part of the cylinder head, but with aluminium alloys a ferrous insert is used. Valve seat inserts have to be used in aluminium cylinder heads, while with cast iron cylinder heads the seats can be induction-hardened. In spark ignition engines running on leaded fuel, the lead compounds lubricate the valve seat, so obviating the need for surface hardening.

8.6 Lubrication and bearings

8.6.1 Lubrication

The frictional energy losses inside an engine arise from the shearing of oil films between the working surfaces. By motoring an engine (the engine is driven without firing) and sequentially dismantling the components it is possible to estimate each frictional component. The results are not truly representative of a firing engine since the cylinder pressures and temperatures are much reduced. None the less, the results in figure 8.13 derived from Whitehouse and Metcalfe (1956) indicate the trends. The frictional losses increase markedly with speed, and at all speeds the frictional losses become increasingly significant at part load operation. Currently much work is being conducted with highly instrumented engines and special test rigs in order to evaluate frictional losses over a range of operating conditions. For example, valve train losses can be derived from strain gauges measuring the torque in the camshaft drive, and piston friction can be deduced from the axial force on a cylinder liner.

The scope for improving engine efficiency by reducing the oil viscosity is limited, since low lubricant viscosity can lead to lubrication problems, high oil consumption and engine wear. The SAE oil viscosity classifications are widely used: there are four categories (5W, 10W, 15W, 20W) defined by viscosity measurements at −18°C (0°F), and a further four categories (20, 30, 40, 50) defined by measurements at 99°C (210°F). Multigrade oils have been developed to satisfy both requirements by adding polymeric additives (viscosity index

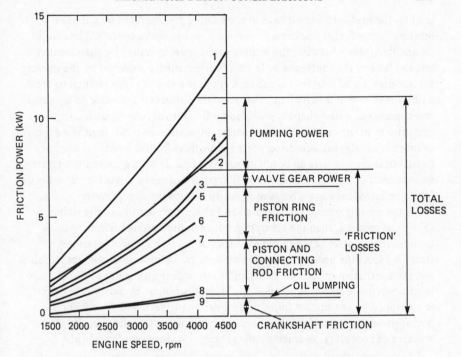

Figure 8.13   Analysis of engine power loss for a 1.5 litre engine with oil
                viscosity of SAE 30 and jacket water temperature of 80°C.
                Curve 1, complete engine; curve 2, complete engine with push
                rods removed; curve 3, cylinder head raised with push rods
                removed; curve 4, as for curve 3 but with push rods in operation;
                curve 5, as for curve 3 but with top piston rings also removed;
                curve 6, as for curve 5 but with second piston rings also removed;
                curve 7, as for curve 6 but with oil control ring also removed;
                curve 8, engine as for curve 3 but with all pistons and connecting-
                rods removed; curve 9, crankshaft only (adapted from
                Blackmore and Thomas (1977))

improvers) that thicken the oil at high temperatures; the designations are thus
SAE 10W/30, SAE 20W/40 etc. Multigrade oils give better cold start fuel
economy because the viscosity of an SAE 20W/40 oil will be less than that of
an SAE 40 oil at ambient conditions.

A disadvantage of the SAE classification is that viscosity measurements are
not made under conditions of high shear. When multigrade oils are subject to
high shear rates the thickening effects of the additives are temporarily reduced.
Thus, any fuel economy results quoted for different oils need a more detailed
specification of the oil; a fuller discussion with results can be found in Blackmore
and Thomas (1977).

There are three lubrication regimes that are important for engine components;
these are shown on a Stribeck diagram in figure 8.14. Hydrodynamic lubrication

is when the load-carrying surfaces of the bearing are separated by a film of lubricant of sufficient thickness to prevent metal-to-metal contact. The flow of oil and its pressure between the bearing surfaces are governed by their motion and the laws of fluid mechanics. The oil film pressure is produced by the moving surface drawing oil into a wedge-shaped zone, at a velocity high enough to create a film pressure that is sufficient to separate the surfaces. In the case of a journal and bearing the wedge shape is provided by the journal running with a slight eccentricity in the bearing. Hydrodynamic lubrication does not require a supply of lubricant under pressure to separate the surfaces (unlike hydrostatic lubrication), but it does require an adequate supply of oil. It is very convenient to use a pressurised oil supply, but since the film pressures are very much greater the oil has to be introduced in such a way as not to disturb the film pressure.

As the bearing pressure is increased and either the viscosity or the sliding velocity is reduced, then the separation between the bearing surfaces reduces until there is contact between the asperities of the two surfaces — point A on figure 8.14. As the bearing separation reduces, the solid-to-solid contact increases and the coefficient of friction rises rapidly, so leading ultimately to the boundary lubrication mode shown in figure 8.15. The transition to boundary lubrication is controlled by the surface finish of the bearing surfaces, and the chemical composition of the lubricant becomes more important than its viscosity. The real area of contact is governed by the geometry of the asperities and the

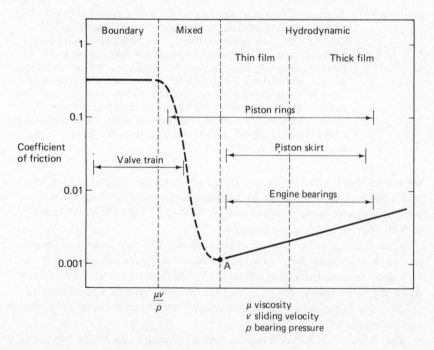

Figure 8.14   Engine lubrication regimes

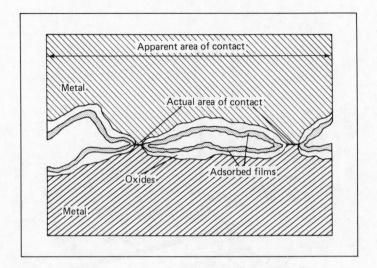

Figure 8.15   Contact between surfaces with boundary lubrication

strength of the contacting surfaces. In choosing bearing materials that have
boundary lubrication, it is essential to choose combinations of material that
will not cold-weld or 'pick-up' when the solid-to-solid contact occurs. The lubricant also convects heat from the bearing surfaces, and there will be additives to
neutralise the effect of acidic combustion products.

Part of an engine lubrication system is shown in figure 8.16; this is the same
engine as is shown in figures 1.10 and 8.7. Oil is drawn from the sump through a
mesh filter and into a multi-lobe positive displacement pump. The pump is
driven by spur gears from the crankshaft. The pump delivery passes through the
oil cooler, and into a filter element. Oil filters should permit the full pump flow,
and incorporate a pressure-relief system, should the filter become blocked.
Filters can remove particles down to 5 μm and smaller.

The main flow from the filter goes to the main oil gallery, and thence to the
seven main bearings. Oil passes to the main bearings through drillings in the
crankshaft; these have been omitted from this diagram for clarity. The oil from
the main bearings passes to the camshaft bearings, and is also sprayed into the
cylinders to assist piston cooling. A flow is also taken from the centre camshaft
bearing to lubricate the valve gear in the cylinder head.

8.6.2 Bearing materials

There are two conflicting sets of requirements for a good bearing material:

(1) the material should have a satisfactory compressive and fatigue strength

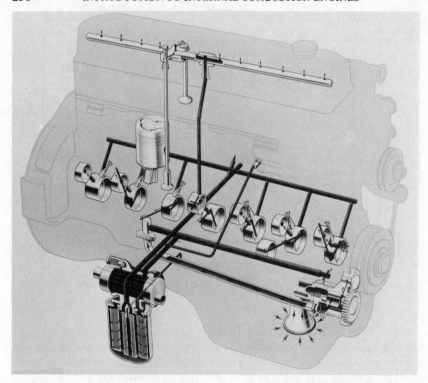

Figure 8.16   Engine lubrication system; see also figures 1.10 and 8.7
(courtesy of Ford)

(2) the material should be soft, with a low modulus of elasticity and a low
melting point.

Soft materials allow foreign particles to be absorbed without damaging the
journal. A low modulus of elasticity enables the bearing to conform readily to
the journal. The low melting point reduces the risk of seizure that could occur in
the boundary regime — all bearings at some stage during start-up will operate in
the boundary regime. These conflicting requirements can be met by the steel-
backed bearings discussed below.

Initially Babbit metal, a tin–antimony–copper alloy, was widely used as a
bearing material in engines. The original composition of Babbit metal or white
metal was 83 per cent tin, 11 per cent antimony and 6 per cent copper. The
hard copper–antimony particles were suspended in a soft copper–tin matrix, to
provide good wear resistance plus conformability and the ability to embed
foreign particles. The disadvantages were the expense of the tin and the poor
high-temperature performance, consequently lead was substituted for tin. The
white metal bearings were originally cast in their housings, or made as thick

shells sometimes with a thick bronze or steel backing. In all cases it was necessary to fit the bearings to the engine and then to hand-scrape the bearing surfaces. This technique made the manufacture and repair of engines a difficult and skilled task.

These problems were overcome by the development of thin-wall or shell-type bearings in the 1930s. Thin-wall bearings are made by casting a thin layer of white metal, typically 0.4 mm thick, on to a steel strip backing about 1.5 mm thick. The manufacture is precise enough to allow the strip to be formed into bearings that are then placed in accurately machined housings. These bearings kept all the good properties of white metal, but gained in strength and fatigue life from the steel backing. To provide bearings for higher loads a lead–bronze alloy was used, but this required hardening of the journals — an expensive process. To overcome this difficulty a three-layer bearing was developed. A thin layer of white metal was cast on top of the lead–bronze lining. To prevent diffusion of this tin into the lead–bronze layer, a plated-nickel barrier was necessary. The expense of three-layer bearings led to the development of single-layer aluminium–tin bearings with up to 20 per cent tin. More recently an 11 per cent silicon–aluminium alloy has been developed for heavily loaded bearings.

The manufacture of such bearings is a specialist task carried out by firms such as GKN Vandervell. Table 8.2 lists some of their bearing materials.

Table 8.2 Properties and composition of bearing materials

Designation and composition		Load-carrying capacity (MN/m^2)	Comments
Lead–bronze VP1			
Lead	14–20%	62	
Tin	4–6%		
Iron	0.5% max.		
Other impurities	0.75% max.		
Copper	Remainder		
Lead–bronze VP2			
Lead	20–26%	48	High-strength bearings with good
Tin	1–2%		fatigue life. Often used with a
Iron	0.5% max.		plated lead–indium overlay
Other impurities	0.75% max.		
Copper	Remainder		
Lead–bronze VP10			
(equivalent to SAE 792)			
Lead	9–11%	82	
Tin	9–11%		

Table 8.2

Designation and composition		Load-carrying capacity (MN/m^2)	Comments
Iron	0.6% max.		
Zinc	0.75% max.		
Antimony	0.3%		
Phosphorus	0.05% max.		
Nickel	0.5%		
Copper	Remainder		
Tin-based Babbitt (also Micro-Babbitt) VP17 (equivalent to SAE 12)			
Tin	88.25% min.	15	Excellent conformability, embed-
Antimony	7.25-7.75%		ability and corrosion resistance
Copper	3.0-3.5%		
Tellurium	0.10-0.14%		
Lead	0.25% max.		
Iron	0.08% max.		
Arsenic	0.10% max.		
Aluminium	0.005% max.		
Other impurities	0.16% max.		
Lead-based Babbitt (also Micro-Babbitt) VP18 (equivalent to SAE 15)			
Antimony	14.5-15.5%	15	Superior strength to VP17 at
Tin	0.9-1.25%		high temperatures. Excellent
Arsenic	0.85-1.15%		conformability, embedability and
Copper	0.5% max.		corrosion resistance
Aluminium	0.005% max.		
Zinc	0.005% max.		
Other impurities	0.25% max.		
Lead	Remainder		
Aluminium-tin VP19			
Tin	18-22%	31	For use in medium-loaded half-shell
Copper	0.7-1.3%		bearings. Excellent corrosion
Iron	0.4% max.		resistance
Silicon	0.4% max.		
Manganese	0.2% max.		
Other impurities	0.35% max.		
Aluminium	Remainder		

8.7 Conclusions

By careful design and choice of materials, modern internal combustion engines are manufactured cheaply with a long reliable life. Many ancillary components such as carburettors, radiators, ignition systems, fuel injection equipment etc. are made by specialist manufacturers. Many engine components such as valves, pistons, bearings etc. are also manufactured by specialists. Consequently, engineers also seek advice from specialist manufacturers during the design stage.

9 Experimental Facilities

9.1 Introduction

The testing of internal combustion engines is an important part of research, development and teaching of the subject. Engine test facilities vary widely. The facilities used for research can have very comprehensive instrumentation, with computer control of the test and computer data acquisition. On the other hand, a more traditional test cell with the engine controlled manually, and the data recorded by the operator, can be better for educational purposes. This second type of test cell is covered in some detail by Greene and Lucas (1969), and is dealt with first in this chapter. The chapter ends with a case study of an advanced engine test system using microprocessor engine control and data acquisition. A final class of test facility not separately discussed here is those used for acceptance tests. Most engines are tested immediately after manufacture to check power output and fuel consumption; the main requirement here is ease of installation.

Before dealing with any test facility it is important to remember that there are certain advantages in using single-cylinder engines for research and development purposes:

(1) No inter-cylinder variation. The manufacturing and assembly tolerances in multi-cylinder engines cause performance differences between cylinders. This is attributable to differences in compression ratio, valve timing etc.
(2) No mixture variation. With fuel injection systems it is difficult to calibrate pumps and injectors to give identical fuel distribution. In carburated engines it is difficult to design the inlet manifold to give the same air/fuel mixture to all cylinders for all operating conditions.
(3) For a given cylinder size the fuel consumption will be less and a smaller capacity (cheaper) dynamometer can be used.

9.2 Simple engine test rig

Figure 9.1 shows the schematic arrangement of a simple engine test rig for a Ruston Oil Engine. This single-cylinder compression ignition engine has the advantage of simplicity and ruggedness. The engine has a bore of 143 mm and a stroke of 267 mm, giving a displacement of 4.29 litres. The engine is governed to operate at 450 rpm, but fine speed adjustment is still necessary. The slow operating speed means that a particularly simple engine indicator can be used to determine the indicated work output of the engine.

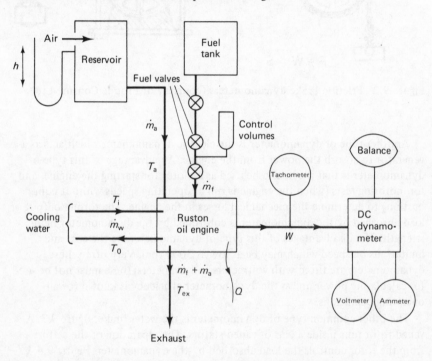

Figure 9.1   Schematic test arrangement for a Ruston Oil Engine

9.2.1 Dynamometers

The dynamometer is perhaps the most important item in the test cell, as it is used to measure the power output of the engine. The term 'brake horse power' (BHP) derives from the simplest form of engine dynamometer, the friction brake. Typically the engine flywheel has a band of friction material around its circumference, and the torque reaction on the friction material corresponds to the torque output of the engine — see figure 9.2.

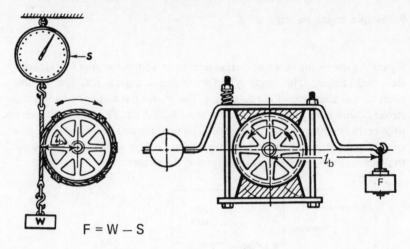

$$F = W - S$$

Figure 9.2   Friction brake dynamometers (courtesy of Froude Consine Ltd)

Another type of dynamometer is the electric dynamometer which acts as a generator to absorb the power from the engine. An advantage of this type of dynamometer is that it can also be used as a motor for starting the engine, and for motoring tests (when the engine is run at operating speeds without combustion) to determine the mechanical losses in the engine. The torque output or load absorbed by the dynamometer is controlled by the dynamometer field strength. The disadvantages of this type of dynamometer are the cost and limitations on speed which may be as low as 3000 rpm. Very often these dynamometers are fitted with voltmeters and ammeters; these must not be used for calculating power unless the dynamometer efficiency is known for all operating conditions.

The other common type of dynamometer is the water brake, figure 9.3. A vaned rotor runs inside a pair of vaned stators. The separation of the stators from the rotor controls the load absorbed by the dynamometer. Figure 9.4 shows typical absorption curves for a hydraulic dynamometer; only engine-operating points between the upper and lower solid lines can be attained. The dashed line shows the typical power output of a diesel engine; in this case only limited powers can be measured below 750 rpm.

Both types of dynamometer measure the load absorbed by the torque reaction on the dynamometer casing. The dynamometer is mounted in bearings co-axial with the shaft, so that the complete dynamometer is free to rotate, but usually within a restricted range, figure 9.5. The torque reaction (T), is equal to the product of the effective lever arm length (l_b) and the net force on the lever arm, F:

$$T = F.l_b \text{ (N m)}$$

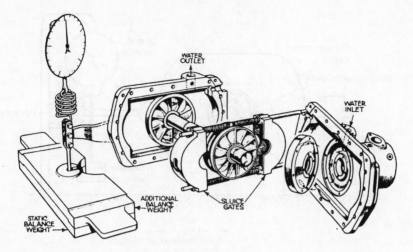

Figure 9.3   Hydraulic dynamometer (courtesy of Froude Consine Ltd)

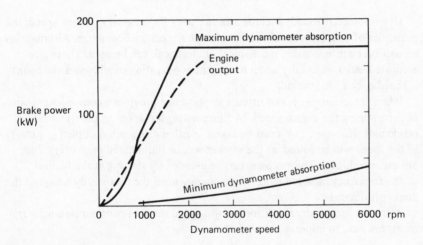

Figure 9.4   Absorption characteristics of a hydraulic dynamometer

The force F must be measured in the datum position for two reasons:

(1) the effective lever arm length becomes $l_b \cos \phi$
(2) away from the datum position a torque will be exerted by the connections for the cable or hose.

Usually a dashpot is connected to the lever arm to damp any oscillations. All these problems are avoided if a load cell is connected to the lever arm, although calibration problems are introduced instead.

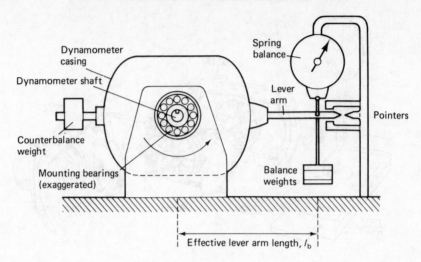

Figure 9.5   Dynamometer mounting system

Dynamometers usually include a tachometer for measuring engine speed; the principles of operation are the same as those for car speedometers. Alternatively, revolution counters, either mechanical or electrical, can be used. These give accurate results, especially when the engine is operating steadily and the count is timed over a long period.

Where a tachometer is not fitted a stroboscope provides a convenient means of determining the engine speed, by illuminating a marker on the flywheel or crankshaft. However, care must be taken or otherwise a submultiple ($\frac{1}{2}$, $\frac{1}{3}$ etc.) of the speed will be found, as the marker will be illuminated once every 2nd, 3rd etc. revolution. This problem can be avoided by starting at the highest strobe frequency, and reducing the frequency until the first steady image of the marker is obtained.

After the measurements of torque and speed the fuel-consumption measurements are next in importance.

9.2.2 Fuel-consumption measurement

The commonest measurement system for fuel consumption is to time the consumption of a fixed volume. This has to be converted to a gravimetric consumption by using the density as determined from a separate test, or deduced from fuel temperature for a specific fuel. A typical arrangement is shown in figure 9.6. In normal operation valve 1 is open and fuel flows directly to the engine. The calibrated volumes are filled when valve 2 is open. If the vent pipe ends below the level of the fuel tank, care is needed in filling the control volumes. To measure the fuel flow rate, valve 1 is closed and any fuel to the engine is drawn from the calibrated volumes; valve 2 must be open. Usually there are a range of

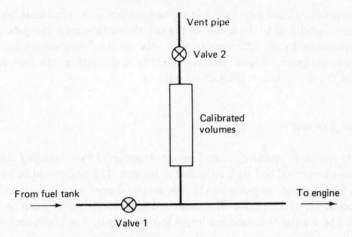

Figure 9.6   Fuel flow measurement

volumes to give the best compromise between accuracy and speed of taking the readings. A common problem is when a vapour-bubble travels back down the fuel line; this obviously invalidates a reading. In a fuel-injection system, if the spill flow is fed back to the tank this must be measured separately; such fuel is much more prone to vapour-bubbles. Although this method gives accurate results , the readings are not instantaneous.

Figure 9.7 shows a flowmeter that does give instantaneous readings, though the accuracy is less than that of the flowmeter previously described. The float

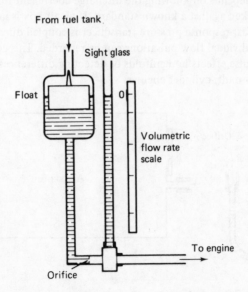

Figure 9.7   Orifice type flowmeter

chamber provides a constant head of fuel, and the pressure drop across the orifice is proportional to the square of the volumetric flow rate. The pressure drop is measured by the difference in head between the float chamber and the sight glass; alongside is a scale calibrated directly in volumetric flow rate. Again, problems can occur with fuel injection systems.

9.2.3 Air flow rate

A simple system to measure the air flow rate is obtained by connecting the air intake to a large rigid box with an orifice at its inlet. The box should be large enough to damp out the pulsations in flow and be free of resonances in the normal speed range of the engine. The pressure drop across the orifice can be measured by a water tube manometer, as shown in figure 9.8. For incompressible flow

$$\dot{m}_a = C_d A_o \sqrt{(2gh\rho_f \rho_a)}$$

where
$\dot{m}_a$ = mass flow rate of air (kg/s)
C_d = discharge coefficient of the orifice
A_o = cross-sectional area of the orifice (m)
g = acceleration due to gravity (m/s^2)
h = height difference between liquid levels in the manometer (m)
ρ_f = density of manometer fluid (kg/m^3)
ρ_a = density of air = $(p/R_a T)$ (kg/m^3).

The accuracy depends on knowing the discharge coefficient for the orifice; this should be checked against a known standard. If an air box is not used and an orifice with a fast-response pressure transducer is coupled directly to the engine intake, then individual flow pulsations can be recorded. This could be useful for studying the pulse effects in manifolds or detecting differences in inlet valve performance in multi-cylinder engines.

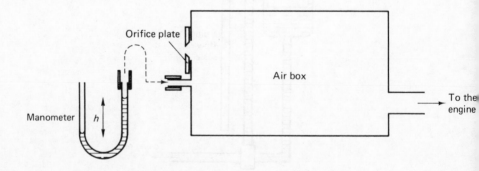

Figure 9.8   Air flow rate tank

An alternative approach is the viscous flowmeter. As no damping is required it is much more compact than an air box. To obtain viscous flow a matrix of passages is used in which the length is much greater than a typical diameter. Since the flow is viscous the pressure drop is proportional to the velocity or volumetric flow rate. The flowmeter has to be calibrated against a known standard.

In general, gas flow rate is more difficult to determine than liquid flow rate. The calibration of liquid flowmeters can always be checked gravimetrically. Air flow rate is needed for calculating both the air/fuel ratio and the volumetric efficiency.

9.2.4 Indicator diagrams

Engine indicators record the pressure/volume history of the engine cylinder contents. The simplest form of engine indicator is the Dobie McInnes mechanical indicator shown in figure 9.9. A piece of paper is attached to a drum, and the rotation of the drum is linked to the piston displacement by a cord wrapped around the drum. The pressure in the cylinder is recorded by a linkage attached to a spring-loaded piston in a cylinder. The indicator cylinder is connected to the engine cylinder by a valve. Unless the indicator cylinder has a much smaller volume than the engine cylinder, the indicator will affect the engine performance.

When the paper is unwrapped from the drum the area of the diagram can be found, and this corresponds to the indicated work per cylinder per cycle. The area can be determined by 'counting squares', by cutting the diagram out and

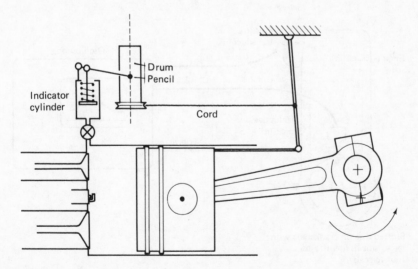

Figure 9.9   Mechanical indicator

weighing it, or by using a planimeter. The planimeter is a mechanical device that computes the area of the diagram by tracing the perimeter. Practice is necessary in order to obtain reliable results, and accuracy can be improved by tracing round the diagram several times. To convert area to work a calibration constant is needed; alternatively imep can be found more directly. The diagram area is divided by its length to give a mean height. When this height is multiplied by the indicator spring constant (bar/mm) the imep can be found directly:

$$\text{imep} = k\, h_d = k\, \frac{A_d}{l_d}$$

where A_d = diagram area
       l_d = diagram length
       h_d = mean height of diagram.

Because of the inertia effects in moving parts — friction, backlash and finite stiffness — mechanical indicators can be used only at speeds of up to about 600 rpm. Also, this simple type of mechanical indicator is not sensitive enough to record the 'pumping losses' during the induction and exhaust strokes.

Electronic systems are now very common for recording indicator diagrams. Care is needed in their interpretation since the pressure is plotted on a time instead of a volumetric basis. As with any electronic equipment, care is also needed in the calibration. The output from the pressure transducer is connected to the y-channel of an oscilloscope and the time base is triggered by an inductive pick-up on the crankshaft — see figure 9.10. The inductive pick-up should also be connected to a second y-channel so that the position of tdc can be accurately recorded. Since tdc occurs during the period of maximum pressure change, a 1°

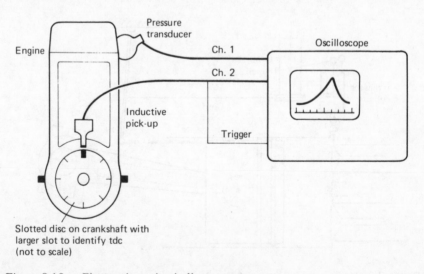

Figure 9.10   Electronic engine indicator system

error in position of tdc can cause a 5 per cent error in imep. The output from the oscilloscope can be recorded photographically. If a transient recorder is connected before the oscilloscope, the output can alternatively be directed to an x–y plotter.

If additional channels are available they can be used to advantage. In spark ignition engines the ignition timing can be recorded by wrapping a wire around the appropriate high-tension ignition lead. For compression ignition engines, injectors with needle lift transducers and fuel pressure transducers can be obtained.

To convert the time base to a piston displacement base it is usual to assume constant angular velocity throughout each revolution. Assuming that the gudgeon pin or 'little-end' is not offset (that is, assuming that the line of motion of the pivot in the piston intersects the axis of rotation of the crankshaft), the piston displacement (x) is given by

$$x = r\left(1 - \cos\theta\right) + (l - \sqrt{(l^2 - r^2 \sin^2\theta)})$$

where θ = crank angle measured from tdc
         r = crank radius (half piston stroke)
         l = connecting-rod length.
When $l \gg r$ the motion becomes simple harmonic.

The pressure transducer requirements are very demanding because of the high temperature and pressures, and the need for a high-frequency response. The transducers usually have a metal diaphragm which is displaced by the pressure. The displacement can be measured, inductively, by capacitance, by a strain gauge or by a piezo-electric crystal. Piezo-electric transducers are common, but have the disadvantage that they respond only to the rate of change in pressure; thus they have to be used in conjunction with a charge amplifier that integrates the signal. The pressure transducer response should be independent of temperature, and the calibration should be free from drift. The pressure transducer should be mounted flush with the cylinder wall or as close to the engine cylinder as possible through a small communicating passage. This minimises the lag in the pressure signal and should avoid introducing any resonances in the connecting passage. For all these reasons it is very difficult to determine imep accurately.

9.2.5 Indicated power

Very often a pressure transducer cannot be readily fitted to an engine, so alternative means of deducing imep are useful. The difference between indicated power and brake power is the power absorbed by friction, and this is often assumed to be dependent only on engine speed. Unfortunately, the friction power also depends on the indicated power since the increased gas pressures cause increases

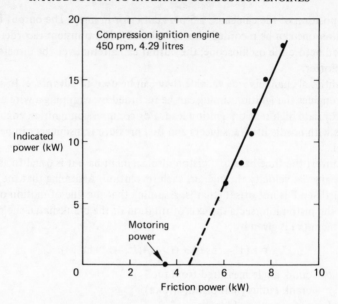

Figure 9.11   Dependence of friction power on indicated power at constant speed

in piston friction etc.; this is shown by figure 9.11. When extrapolated to zero-indicated power, the friction power is about 4.5 kW. This can be compared to 3.5 kW motoring power — the power output of the electric dynamometer turning the engine at the same speed.

If friction power is assumed to be independent of indicated power, then the friction power can be deduced from the Morse test. This is applicable only to multi-cylinder engines (either spark or compression ignition), as each cylinder is disabled in turn. When a cylinder is disabled the load is reduced so that the engine returns to the test speed; the reduction in power corresponds to the indicated power of that cylinder.

For a n-cylinder engine

$$\sum^{n} \text{indicated power} - \sum^{n} \text{friction power} = (\text{brake power})_n$$

With one-cylinder disabled

$$\sum^{n-1} \text{indicated power} - \sum^{n} \text{friction power} = (\text{brake power})_{n-1}$$

Subtracting:

indicated power of disabled cylinder = reduction in brake power

This underestimates the friction power since the disabled cylinder also has reduced friction power. However, the test does check that each cylinder has the same power output.

A method for estimating the friction power of compression ignition engines is Willan's line. Again it is assumed that at constant speed the friction power is independent of indicated power, but in addition it is assumed that the indicated efficiency is constant. This is a reasonable assumption away from maximum power. Figure 9.12 shows a plot of fuel consumption against power output. Willan's line is when the graph is extrapolated to zero fuel flow rate to determine friction power. Figure 9.12 is for the same engine as shown in figure 9.11, but Willan's line suggests a friction power of only 2.5 kW.

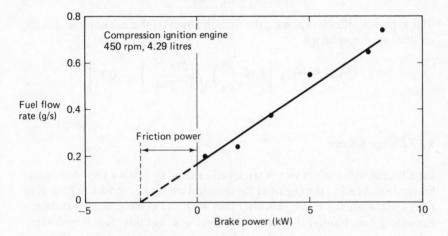

Figure 9.12   Willan's line for a Diesel engine

9.2.6 Engine test conditions

Various standards authorities (BS, DIN, SAE) are involved with specifying the test conditions for engines, and how allowance can be made for variations in ambient conditions.

In the past a wide range of performance figures could be quoted for a given engine, depending on which standard was adopted and how many of the engine ancillary components were being driven (water pump, fan, alternator etc.). Obviously it is essential to quote the standard being used, and to adhere to it.

Corrections for datum conditions vary, and in general they are more involved for compression ignition engines, whether turbocharged or naturally aspirated. Corrections for spark ignition engines in the *SAE Handbook* are as follows:

$$\text{ambient test conditions} \begin{cases} 95 < p < 101 \text{ kN/m}^2 \\ 15 < T < 43°\text{C} \end{cases}$$
should be in the range

where p is ambient pressure and T is ambient temperature.

The corrections are applied to indicated power, where $(\dot{W_i})_o$ is the observed value and $(\dot{W_i})_c$ is the corrected value:

$$(\dot{W_i})_c = (\dot{W_i})_o \left(\frac{99}{p_d}\right) \sqrt{\left(\frac{T + 273}{298}\right)}$$

where $p_d = p - p'_{water}$, the partial pressure of dry air (kN/m^2)

Values for brake power $(\dot{W_b})$ are found from a knowledge of the friction power $(\dot{W_f})$:

$$\dot{W_i} = \dot{W_b} + \dot{W_f}$$

This approach relies on knowing the friction power. If this has not been found, an alternative approach is

$$(\dot{W_b})_c = (\dot{W_b})_o \left[1.18\left(\frac{99}{p_d}\right) \sqrt{\left(\frac{T + 273}{298}\right)} - 0.18\right]$$

9.2.7 Energy balance

Experiments with engines very often involve an energy balance on the engine. Energy is supplied to the engine as the chemical energy of the fuel and leaves as energy in the cooling water, exhaust, brake work and extraneous heat transfer. Extraneous heat transfer if often termed 'heat loss', but this usage is misleading as heat is energy in transit and the 1st Law of Thermodynamics states that energy is conserved.

The heat transfer to the cooling water is found from the temperature rise in the coolant as it passes through the engine and the mass flow rate of coolant. The temperature rise is most commonly measured with mercury in glass thermometers. The mass flow rate of coolant is usually derived from the volumetric flow rate. Common flow-measuring devices include tanks, weirs and variable-area flowmeters such as the Rotameter. The Rotameter has a vertical-upwards flow through a diverging graduated tube; a float rises to an equilibrium position to indicate the flow rate. To conserve water the coolant is usually pumped in a loop with some form of heat exchanger. The heat exchanger should be regulated to control the maximum engine-operating temperature. The engine coolant flow can be adjusted to make the temperature rise sufficiently large to be measured accurately without making the flow rate too small to be measured accurately.

The energy leaving in the exhaust is more difficult to determine. If the gas temperature is measured, the mean specific heat capacity can be estimated in order to calculate the enthalpy in the exhaust. Sometimes a known flow rate of water is sprayed into the exhaust, and the temperature is measured after the water has evaporated. The exhaust can include unburnt fuel, notably with spark ignition engines operating on rich mixtures; this can make a significant difference to an energy balance.

Heat transfer from the engine cannot be readily determined from temperature measurements of the engine. If the engine is totally enclosed, the temperature rise and mass flow rate of the cooling air can be used to determine the heat transfer.

Finally, brake power should be used in the energy balance, not indicated power. The power dissipated in overcoming friction degenerates to heat, and this is accounted for already.

9.3 Experimental accuracy

Whenever an experimental reading is taken there is an error associated with that reading. Indeed, it can be argued that any reading is meaningless unless it is qualified by a statement of accuracy. There are three main sources of error:

(1) the instrument is not measuring what is intended
(2) the instrument calibration is inaccurate
(3) the instrument output is incorrectly recorded by the observer.

To illustrate the different errors, consider a thermocouple measuring the exhaust gas temperature of an engine. Firstly, the temperature of the thermocouple may not be the temperature of the gas. Heat is transferred to the thermocouple by convection, and is transferred from the thermocouple by radiation, and to a lesser extent by conduction along the wires. If the gas stream has a temperature of 600°C and the pipe temperature is about 400°C, then the thermocouple could give a reading that is 25 K low because of radiation losses (Rogers and Mayhew (1980a)).

Secondly, there will be calibration errors. Thermocouple outputs are typically $40 \mu V/K$ and very close to being linear; the actual outputs are tabulated as functions of temperature for different thermocouple combinations. The output is very small, so amplification is often needed; this can introduce errors of gain and offset that will vary with time. Thermocouples also require a reference or cold junction, and this adds scope for further error whether it is provided electronically, or with an additional thermocouple junction in a water/ice mixture. The output has to be indicated on some form of meter, either analogue or digital, with yet further scope for errors.

Thirdly, errors can arise through misreading the meter; this is less likely with digital meters than with analogue meters.

In serious experimental work the instrumentation has to be checked regularly against known standards to determine its accuracy. Where this is not possible, estimates have to be made of the accuracy; this is easier for analogue instruments than for digital instruments. In a well-designed analogue instrument (such as a spring balance, a mercury in glass thermometer etc.), the scale will be devised so that full benefit can be obtained from the instrument's accuracy. In other words, if the scale of a spring balance has 1 Newton divisions and these can be sub-divided into quarters, then it is reasonable to do so, and to assume that the accuracy is also $\pm\frac{1}{4}$ Newton. Of course a good spring balance would have some indication of its accuracy engraved on the scale.

This approach obviously cannot be applied to digital instruments. It is very tempting, but wrong, to assume that an instrument with a four-digit display is accurate to four significant figures. The cost of providing an extra digit is much less than that of improving the accuracy of the instrument by an order of magni-tude! For example, most thermocouples with a digital display will be accurate to only one degree.

There are many books, such as Adams (1975), that deal with instrumentation and the handling of results. One possible treatment of errors uses binomial approximations. If a quantity u is dependent on the quantities x, y and z such that

$$u = x^a y^b z^c$$

then for sufficiently small errors

$$\frac{\delta u}{u} \approx a\left(\frac{\delta x}{x}\right) + b\left(\frac{\delta y}{y}\right) + c\left(\frac{\delta z}{z}\right)$$

where δu denotes the error associated with u etc.
As an example consider equation (7.5):

$$T_{2s} = T_1 \left(\frac{p_2}{p_1}\right)^{(\gamma-1)/\gamma}$$

If the error in T_1 is $\pm\frac{1}{2}$ per cent and the error associated with the pressure ratio (p_2/p_1) is ± 5 per cent, then the error associated with the isentropic compressor temperature T_{2s} is

$$\pm\frac{1}{2} \pm \frac{1.4 - 1}{1.4} \times 5 \text{ per cent}$$

which is ± 1.9 per cent.
This is a pessimistic estimate of the error, since it assumes a worst possible com-bination of errors that is statistically unlikely to occur.

Sometimes it is possible for the experimenter to minimise the effect of errors. Consider the heat flow to the engine coolant:

$$\dot{Q} = \dot{m} c_p (T_{out} - T_{in})$$

Suppose the mass flow rate ($\dot{m}$) of coolant is 4.5 kg/s and the inlet and outlet temperatures are 73.2°C and 81.4°C, respectively. If the errors associated with mass flow rate are ±0.05 kg/s and the errors associated with the temperature are ± 0.2 K, then

$$Q = (4.5 \pm 0.05)c_p [(81.4 \pm 0.2) - (73.2 \pm 0.2)]$$
$$= (4.5 \pm 0.05)c_p(8.2 \pm 0.4)$$
$$= (4.5 \pm 1.11 \text{ per cent})c_p(8.2 \pm 4.88 \text{ per cent})$$
$$= 4.5 \times c_p \times 8.2 \pm 6 \text{ per cent}$$

It can be readily shown by calculus that the errors would be minimised if the percentage error in each term were equal. Denoting the optimised values of mass flow rate as $\dot{m}'$ and temperature difference as $\Delta T'$, then

$$\frac{0.05}{\dot{m}'} = \frac{0.4}{\Delta T'}, \quad \Delta T' = 8\dot{m}'$$

and for the same heat flow

$$\dot{m}'.\Delta T' = 4.5 \times 8.2$$

Combining

$$\dot{m}' = \sqrt{\left(\frac{4.5 \times 8.2}{8}\right)}$$
$$= 2.15 \text{ kg/s}$$
$$\text{and } \Delta T' = 17.18$$

The error is now reduced to

$$\pm \left(\frac{0.05}{2.15}\right) \pm \left(\frac{0.4}{17.18}\right)$$

or ± 4.65 per cent.

Since the effect of errors is cumulative, always identify the weakest link in the measurement chain, and see if it is possible to make an improvement.

9.4 Advanced test systems

Engine test cells are becoming increasingly complex for several reasons. Additional instrumentation such as exhaust gas analysis has become necessary and, in the search for the smaller gains in fuel economy, greater accuracy is also necessary. Consequently the cost of engine test cells has escalated, but fortunately the cost of computing equipment has fallen.

Computers can be used for the control of a test and data acquisition, thus improving the efficiency of engine testing. The computer can also process all the data, carry out statistical analyses, and plot all the results. The design of the test facility and computer system will depend on its use — whether it is for research, development, endurance running or production testing. There are, of course, overlaps in these areas, but the facility described here will be for development work.

Descriptions of computer-controlled test facilities are given by Watson *et al.* (1981) and Donnelly *et al.* (1981). Complete facilities are marketed by firms such as Froude Consine and Schenck; a Schenck system is described here. A block diagram of a computerised test facility is given in figure 9.13. Separate microprocessors (micros) are used for data acquisition and test point control. A single microprocessor could be used but that would reduce the data-acquisition rate, and reduce the storage space for data and test cycles. The printer and VDU are linked to the micros so that the test pattern can be chosen and then monitored. The host computer can be linked to many such microsystems, and it provides a more powerful data-processing system with greater storage. Data from many tests can then be archived on magnetic discs or tape. The host computer can also provide sophisticated graphics facilities and plotters that would be under-used if dedicated to a single test cell.

A typical requirement is to produce an engine fuel-consumption map. This requires running the engine over a wide range of discrete test points. Each test point is specified in terms of speed and load (or throttle position), and the sequence and duration of each test point is stored in the micro. The software for setting up a test programme is stored in an EPROM (erasable programmable read only memory) and is designed so that new or existing test programmes can be readily used. At each test point the dynamometer controller and throttle controller work in conjunction to obtain the desired test condition. Additional parameters such as ignition timing and air/fuel mixture can also be controlled. At each test point a command is sent to the data-acquisition system to sweep all the channels (that is, to take measurements from all the specified transducers). At the end of the test the data can be sent for analysis on the host computer. At each test point, commands can be sent to instrumentation that does not take continuous readings; for instance, the initiation of a fuel-consumption measurement.

The data-acquisition system also monitors values from the instrumentation (such as engine speed, water temperature) and if the values fall outside a certain range then an alarm can be set. If the values fall outside a specified wider range, then the engine can be stopped, and a record kept of what caused the shutdown. The software also permits the separate calibration of transducer channels. The transducer inputs have to be linear and the calibration is defined in terms of specified inputs.

Other parameters such as oil and water temperature normally have separate control systems. These would apply closed loop control to the flow through the appropriate heat exchanger.

In test cells for production testing it is usual to have a pallet-mounting system for the engine, in order to minimise the cell downtime. Such systems may also be justified in development testing where many different engines are being tested.

The additional instrumentation that might be found in a sophisticated test cell includes exhaust gas analysis equipment, and equipment for measuring the ignition delay in diesel engines. There are three main types of exhaust gas analysis equipment, using infra-red absorption, flame ionisation detection and gas chromatography.

Non-dispersive infra-red instruments are used to measure the concentrations of carbon dioxide, carbon monoxide and nitric oxide; the technique is not accurate for hydrocarbons. A sample of gas in a test cell is exposed to infra-red radiation. Certain characteristic wavelengths are absorbed by molecular vibration: 4.0–4.5 μm for carbon dioxide, 4.5–5.0 μm for carbon monoxide. The instrument is calibrated by gas mixtures of known composition.

Flame ionisation detection is primarily used for measuring the hydrocarbon concentrations in exhaust samples. The exhaust sample is burnt in a hydrogen flame and the degree of flame ionisation is measured by the electron flow from a cathode to an anode in the flame. The hydrogen itself produces very little ionisation, so the ionisation current is directly proportional to the number of carbon atoms in the mixture. Again the instrument is calibrated from samples of known concentration.

Gas chromatography is used for identifying particular hydrocarbon species in the exhaust gas. The gas sample and carrier gas are passed into the column – a long tube that contains a medium (liquid or solid) that tends to absorb the constituents in the sample. Since different molecules pass through the column at different rates, when the carrier gas leaves the column it contains the different molecules in discrete groups, and the different molecules can be identified by their residence time. The exit of the species from the column is detected by measuring the ionisation in a flame. The chromatograph is calibrated by injecting samples of known gases into the carrier gas.

Smoke measurements are particularly important for compression ignition engines. There are two main systems: either the absorption of light or the change in reflectance of a white filter paper is measured (the Bosch smoke test).

Ignition delay is a very important measure of the quality of diesel fuel. However, there are different measurement systems and they give different results. Ignition delay is the time between the start of fuel injection and the start of combustion. The start of fuel injection can be deduced from the fuel-line pressure, or more accurately from an injector needle lift transducer. Fuel-line pressure can be measured from strain gauges on the fuel line; needle lift measurement is more difficult. Needle lift transducers have a coil in the injector; this is connected to a circuit resonating at a high frequency (typically 2 MHz). When the injector needle lifts off its seat the inductance of the coil changes, and the resonant circuit becomes detuned.

The start of ignition is more difficult to determine. Ignition can be sensed directly by photo-sensitive devices or deduced from the cylinder pressure record. One accepted technique is to evaluate the second derivative of pressure with respect to time, and to look for a transition. However, sophisticated electronic filtering is required to eliminate spurious signals.

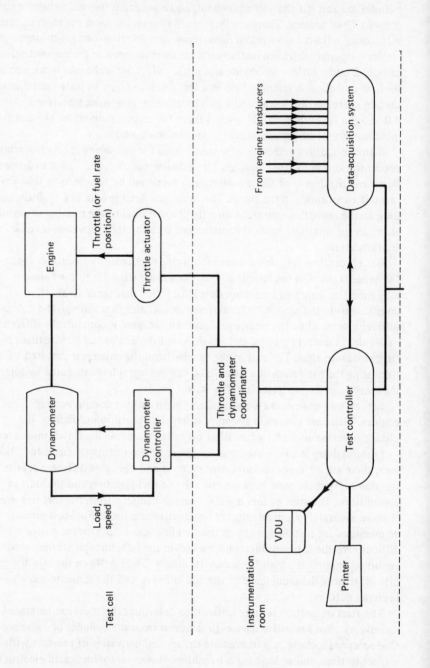

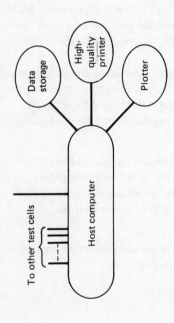

Figure 9.13 Computer-based test facility

9.5 Conclusions

Engine testing is an important aspect of internal combustion engines, as it leads to a better understanding of engine operation. This is true whether the engine is a teaching experiment or part of an engine-development programme. As in any experiment, it is important to consider the accuracy of the results. The first decision is the accuracy level that is required; too high a level is expensive in both time and equipment. The second decision is to assess the accuracy of a given test system; this is of ever-increasing difficulty owing to the rising sophistication of the test equipment.

The nature of development testing is also changing. The use of computer control and data acquisition have been complemented by increasing levels of engine instrumentation. This leads to large quantities of data, and a need for effective post-processing. Needless to say, such systems are expensive, but the need for their use is held back by two factors:

(1) The number of different engines produced is reduced by manufacturers standardising on engine ranges, and by collaboration between companies.
(2) The decreasing cost of computing time, and the use of increasingly powerful computer models, leads to greater optimisation prior to the start of development testing.

However, these factors are balanced by increasing restrictions on engine emissions, and the ever-rising difficulty of improving engine fuel economy.

10 Case Studies

10.1 Introduction

The three engines that have been chosen as case studies are the Jaguar V12 HE spark ignition engine, the Chrysler 2.2 litre spark ignition engine, and the Ford (high-speed) 2.5 litre DI Diesel engine. Each engine has been chosen because of its topicality, and characteristics that are likely to be seen also in subsequent engines. The Jaguar engine uses a May combustion chamber; this permits the use of a high compression ratio and the combustion of lean mixtures, both of which lead to economical operation. The Chrysler 2.2 litre spark ignition engine is typical of current practice in the USA and thus has low emissions of carbon monoxide, unburnt hydrocarbons and oxides of nitrogen.

The Ford compression ignition engine achieves short combustion times and thus high engine speeds by meticulous matching of the air motion and fuel injection. By utilising direct injection, as opposed to indirect injection into a pre-chamber, the pressure drop and heat transfer in the throat to the pre-chamber are eliminated. This immediately leads to a 10–15 per cent improvement in economy and better cold starting performance.

10.2 Jaguar V12 HE engine

10.2.1 Background

The design and development of the original version of the Jaguar V12 engine are described by Mundy (1972), in a paper published at the time of the engine's introduction. In 1981 the HE (high-efficiency) version was introduced with the May combustion chamber. The different compression ratios that have been used are shown in table 10.1.

Table 10.1 Compression ratios for different engine builds

		Compression ratios	
Market	Fuel	V12 (1972)	V12 HE (1981)
Europe	97 RON	9:1	12.5:1
USA	91 RON (lead free)	7.8:1	11.5:1

The V12 engine was designed as an alternative to a six-cylinder in-line engine that had originated some 25 years earlier. An engine capacity of about 5 litres was needed, and the trial engine had to fit into the same space as the six-cylinder engine. While a V8 engine would have been feasible, a V12 engine has complete freedom from all primary and secondary forces and moments, as well as closer firing intervals. Figure 10.1 shows the relative smoothness of six-cylinder, eight-cylinder and twelve-cylinder engines. Marketing considerations also favoured a V12 engine: Jaguar would be the only volume producer of a V12 engine, while V8 engines are quite common in the USA. Another consideration was that, for competition use, a 5 litre V12 engine with a short stroke (70 mm) would permit ultimate power output since the engine would be able to run safely at 8000–8500 rpm.

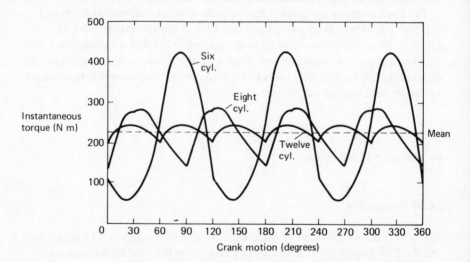

Figure 10.1   Torque characteristics of 6-cylinder, 8-cylinder and 12-cylinder engines (from Campbell (1978))

10.2.2 Engine development

Initial development work was with the classic arrangement of twin overhead camshafts and hemispherical combustion chambers, as had been used on the six-cylinder XK engine. The final production engine had a single overhead camshaft, and this was a consequence of extensive single-cylinder engine tests, and removing the requirement for a competition engine. The outputs of different engine configurations are shown in figure 10.2. The advantages of the single overhead camshaft arrangement were:

  (i) it was lighter, cheaper, and easier to manufacture
 (ii) it had simpler and quieter camshaft drives
(iii) it was easier to install the final engine
 (iv) it had significantly better fuel consumption
  (v) it gave better performance below about 5000 rpm.

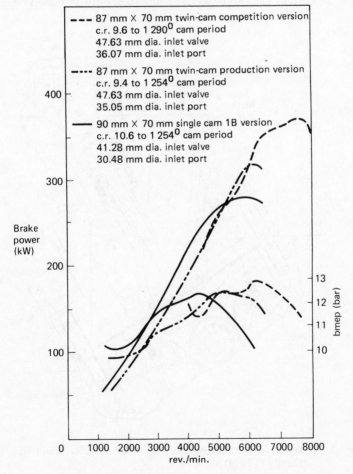

Figure 10.2   Power curves of single and twin cam full-scale engines (from Mundy (1972))

The single overhead camshaft cylinder head is shown in figure 10.3, and there are many similarities with figure 6.2. To minimise the weight of the engine, alloy castings are used for the cylinder head, block and sump. The alloy cylinder head necessitates the use of valve guides and sintered iron valve seat inserts. The camshaft is mounted directly over the valve stems and the cam follower is an inverted bucket-type tappet. The valve clearances are adjusted by interchangeable hardened steel shims placed between the valve stem and the tappet. This so-called Ballot type of valve train provides a very stiff valve gear that is highly suited to high-speed operation. The main disadvantage is that the camshaft has to be removed in order to adjust the valve clearances; however, adjustment is needed much less frequently than for many other arrangements.

The design of the induction and exhaust passages, and even the valve seat inserts, were the subjects of much development work. The cylinder block is also alloy, and the cuff-type liners, which are like wet liners at the top but like dry liners at the bottom, can also be seen in figure 10.3. Cuff-type liners eliminate

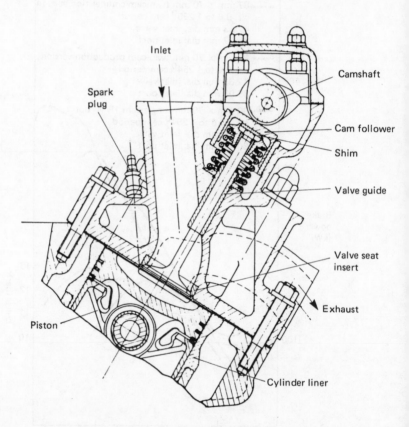

Figure 10.3   Jaguar V12 single overhead camshaft cylinder head (from Mundy (1972))

possible sealing problems associated with the base of wet liners, provide better heat transfer than dry liners, and also have reduced differential thermal expansion effects that would otherwise influence the cylinder head clamping loads. Thermal expansion control slots in the pistons can also be seen in figure 10.3.

Various cylinder block arrangements were considered, before adopting the open-deck alloy cylinder block shown in figure 10.4. The advantage of an open-deck design is that it enables the block to be die-cast; the disadvantage is the reduction in torsional stiffness. By using a deep skirt to the cylinder block (that is, by forming the joint to the sump 100 mm below the crankshaft centre-line) the torsional stiffness of an open-deck engine was greater than that of an alternative design with a closed deck. A cast iron cylinder block was also tested; it weighed 50 kg more than an alloy cylinder block and did not produce a detectably quieter engine. The crankshaft is made from a Tufftrided manganese-molybdenum steel.

Figure 10.4   V12 cylinder block from the top, showing the cylinder liners and the timing chains (from Mundy (1972))

The crescent-type oil pump is mounted directly on the crankshaft, in the space created at the front of the crankcase by the offset between the two banks of cylinders. With any fixed displacement pump driven directly by the engine, the pump delivery is too great at high speeds if the pump displacement is sufficient to provide adequate supply at low speeds. This characteristic is turned to advantage by using the flow from the relief valve to divert through an oil cooler. Thus at high engine speeds when oil cooling is most needed, the flow will be highest through the oil cooler. The oil-cooler circuit is shown in figure 10.5. The oil flow, directed by the relief valve, passes through the oil cooler and back to the pump inlet. This system minimises the flow from the oil pick-up, and also minimises the oil temperature in the pump. The oil is cooled by water that comes direct from the radiator, before entering the water pump. Figure 10.6 shows the oil flow as a function of engine speed. The pump delivery increases almost linearly with engine speed (thus showing almost no dependence on pressure), and the flow is unaffected by the oil viscosity. The oil flow to the engine also increases with engine speed, but with high oil temperatures the reduced viscosity increases the oil demand. The difference between these two flows is the relief flow that passes through the oil cooler.

10.2.3 Jaguar V12 ignition and mixture preparation development

During the initial development, twin six-cylinder ignition distributors were used. These were necessary to ensure adequate spark energy with the very high spark rates from a mechanically operated contact breaker. One distributor incorporated two sets of contact breakers, with the usual vacuum advance and centrifugal advance mechanisms. Even so, it was difficult to match the settings of the contact breakers, and this led to unacceptable variations in ignition timing, especially at higher speeds.

Fortunately, by the time the engine was introduced the Lucas Opus (oscillating pick-up) ignition system was available with a suitable twelve-cylinder distributor. In the Opus ignition system the distributor cam is replaced by a timing rotor with ferrite rod inserts, and the contact breaker is replaced by a pick-up module. An oscillating signal is sent to the primary winding in the pick-up, and a small voltage is induced in the secondary winding. When a ferrite rod passes the pick-up, the coupling between the two coils increases, and a much larger voltage is induced in the pick-up secondary winding. The signal from the secondary winding is used to switch transistors that control the current to the primary winding of the coil.

When the May combustion chamber was introduced, a higher-energy ignition system was needed, and the Lucas constant-energy ignition system was adopted. This system has a variable-reluctance pick-up; a coil is wound around a magnet, and when the 'reluctor' (a toothed cam) changes the magnetic circuit a voltage

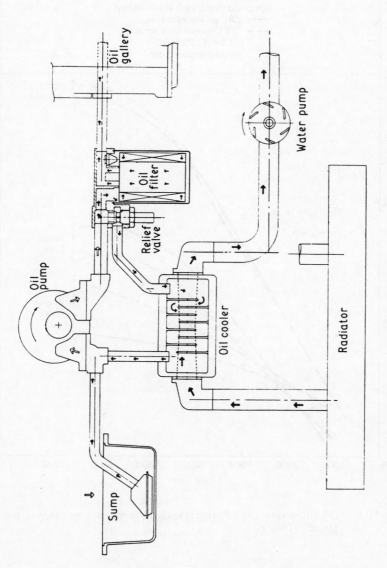

Figure 10.5   Diagrammatic layout of the oil cooler circuit (from Mundy (1972))

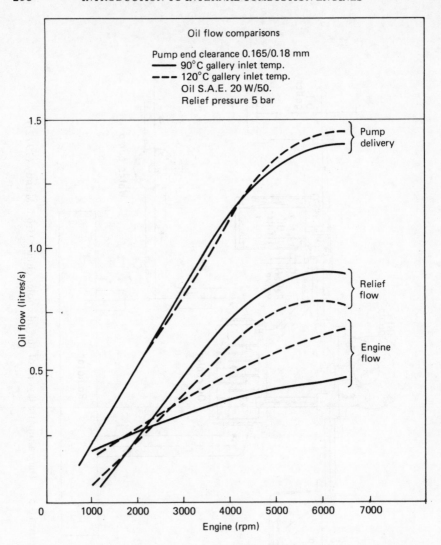

Figure 10.6   Oil pump flow characteristics with oil temperature changes (from Mundy (1972))

is induced in the pick-up coil. The primary coil current of 8 amps is then controlled by transistor switching circuits.

Development work proceeded with both carburettor and fuel injection systems. Mechanically controlled fuel injection systems gave good maximum power and torque, but poor engine emissions. Electronically controlled fuel injection systems that were being developed were not initially available, and the engine was introduced with four variable venturi carburettors. Electronic fuel injection (by Lucas, in conjunction with Bosch) was finally introduced in 1975.

10.2.4 Combustion chamber development

For a single overhead camshaft there are distinct production advantages in having a flat cylinder head, and combustion chambers in the piston. The first piston crown design was of true Heron form: a shallow bowl in the piston of about two-thirds of the bore diameter and with a large squish area. The clearance in the squish area at top dead centre was about 1.25 mm with cut-outs to clear each valve. This is the arrangement shown in figure 10.3. In production, the squish clearance was increased to 3.75 mm, with a shallower bowl in the piston. This arrangement increased the power output, reduced the emissions and did not require cut-outs to clear the valves.

Initially the engine ran satisfactorily on 99 RON fuel with a compression ratio of 10.6:1; this was reduced to 10:1 and finally to 9.0:1, to reduce exhaust emissions. The lower compression ratio reduces the in-cylinder temperatures and pressures, and this reduces the formation of carbon monoxide and nitrogen oxides. For lead-free fuel (91 RON) the compression ratio was reduced to 7.8:1; the effect of compression ratio on engine output is shown in figure 10.7.

In 1981 the Jaguar V12 HE (high-efficiency) engine was introduced with a May combustion chamber — see figure 10.8. The piston is flat topped and this minimises heat transfer to the piston. The combustion chamber is fully machined in the cylinder head, since casting tolerances would give unacceptable variations in the combustion chamber geometry and the compression ratio. The May combustion chamber has a circular disc-shaped recess around the inlet valve, and a connecting passage that is tangential to a much larger recess around the exhaust valve that also contains the spark plug. The swirl generated by the tangential passage, and the squish generated by the flat-topped piston, ensure rapid and controlled combustion. Prior to ignition, the charge helps to cool the exhaust valve, and after ignition the very compact combustion chamber around the exhaust valve produces rapid combustion to reduce the risk of knock. Furthermore, the compact combustion chamber concentrates the charge so that combustion of a lean mixture can be self-sustaining. The result is a combustion chamber that permits the engine to use a 12.5:1 compression ratio with 97 RON fuel (11.5:1 with 91 RON fuel in the USA), and to burn mixtures with an air/fuel ratio as lean as 23:1. The cooler-running piston allows the top-land to be reduced, and this reduces the quench areas and unburnt hydrocarbon emissions. Low CO and NO_x are also inherent features of the V12 HE engine (Crisp (1984)).

The original objective was to produce an engine with comparable power output; the success in achieving this target is shown by figure 10.9. Furthermore, the maximum torque occurred at a lower speed; this permitted a change in the vehicle final drive ratio. The combination of more efficient engine and higher

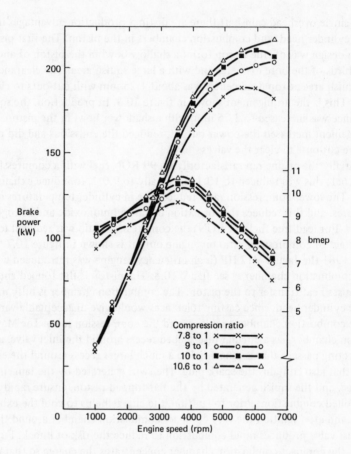

Figure 10.7   Effect of compression ratio on Jaguar V12 engine performance
(from Mundy (1972))

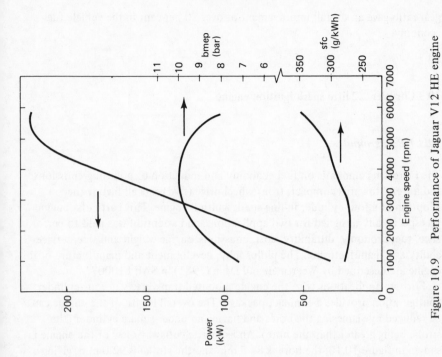

Figure 10.9   Performance of Jaguar V12 HE engine

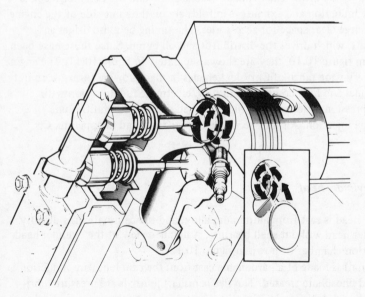

Figure 10.8   Jaguar V12 HE combustion chamber. Insert: underside view of the swirl pattern (with acknowledgement to Newton *et al.* (1983))

gear ratio gave an overall improvement of over 20 per cent in the vehicle fuel economy.

10.3 Chrysler 2.2 litre spark ignition engine

10.3.1 Background

The increasing emphasis on fuel economy and reduction of polluting emissions, and a trend towards compact, front wheel drive (fwd) cars all favour the adoption of a four-cylinder in-line spark ignition engine. This particular engine was specifically designed for fwd applications, and attention was paid to performance, fuel economy, durability, cost, emissions, engine weight and size, serviceability and manufacturing. The philosophy, development and manufacture of this engine are described by Weertman and Dean (1981) in SAE 810007.

For fwd applications with the engine mounted transversely, a four-cylinder configuration provides a compact package. The overall length of the engine can be reduced by siamesing the bores and having an under-square cylinder (the stroke being greater than the bore). An isometric cut-away view of the engine is shown in figure 10.10; the bore is 87.5 mm and the stroke is 92.0 mm, giving a swept volume of 2.213 litres.

The single overhead camshaft is driven by a toothed belt, and operates the in-line valves via rocker arms. The cylinder head is alloy, and the cylinder block is cast iron; the induction and exhaust manifolds are on the same side of the engine to provide a large clear space for accessories. The timing belt also drives an accessory shaft, which drives the distributor and oil pump. Since these have been removed from figure 10.10, they are shown separately in figure 10.11. The engine is used in a very wide range of Chrysler vehicles in the following forms: carburetted with normal and high compression ratios, electronic fuel injection with normal compression ratio, and as a turbocharged engine with multi-point electronic fuel injection. All the specifications can be found in appendix C, section C.7.

10.3.2 The cylinder head

The cylinder head is cast from aluminium alloy, and the camshaft runs directly in the cylinder head without shell bearings. A detailed view of the cylinder head and combustion chamber is shown in figure 10.12.

The camshaft is made of a hardenable cast iron; the cam lobes are induction hardened and phosphate treated. The five camshaft journals are pressure lubricated, and 0.8 mm drilled holes also direct an oil jet on to the cam-lobe/rocker-

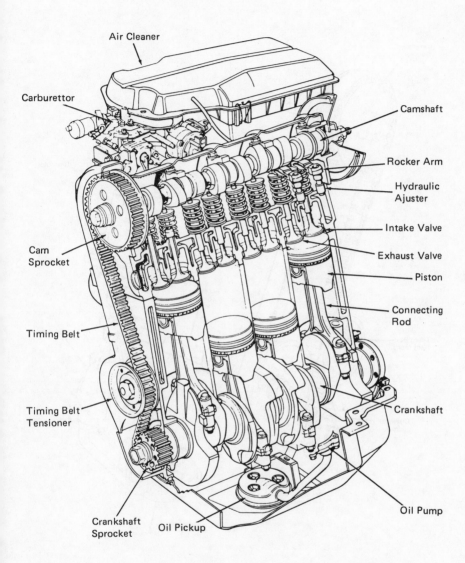

Figure 10.10   Chrysler 2.2 litre four-cylinder engine — longitudinal section
(reprinted with permission, © 1981 Society of Automotive
Engineers, Inc.)

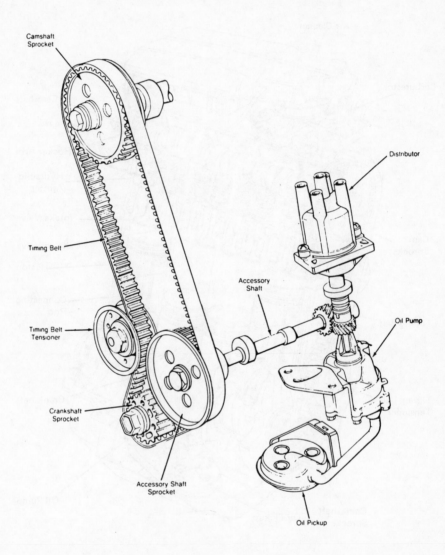

Figure 10.11   Chrysler 2.2 litre four-cylinder engine: timing belt drive −
accessory shaft − distributor and oil pump (reprinted with
permission, © 1981 Society of Automotive Engineers, Inc.)

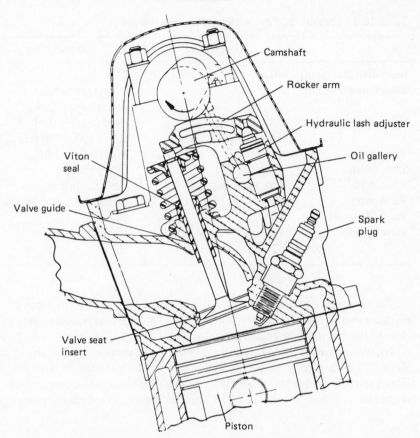

Figure 10.12   Chrysler 2.2 litre engine cylinder head (reprinted with permission,
                © 1981 Society of Automotive Engineers, Inc.)

arm contact area. The rocker arm is cast iron, and pivots on a hydraulic lash
adjuster (hydraulic tappet). The tappet adjustment range is 4.5 mm, and the
valve geometry is such that the valves can never hit the piston.

The valves and associated components have also been designed for long life.
The valve spring rests on a steel washer to prevent the spring embedding in the
cylinder head. The Viton (a heat-resistant synthetic rubber) seals on the valve
spindles prevent oil entering the combustion chamber through the valve guides;
this could be a problem with the copious supply of lubricant to the overhead
camshaft. Table 10.2 highlights some of the design and materials aspects of the
valve gear.

The valve guides and seat inserts are all inserted after pre-cooling in liquid
nitrogen.

Table 10.2 Chrysler 2.2 litre engine valve components

	Inlet valve	Exhaust valve
Head diameter (mm)	40.6	35.4
Stem diameter (mm)	8	8 (tapered to allow for thermal expansion)
Valve material	SAE 1541 steel with hardened tip	Head: 21–2N steel Stem: SAE 4140 steel with hardened tip
Stem finish	Flash chrome	Heavy chrome
Valve guide	Low alloy cast iron	Hardenable iron
Valve seat	Sintered copper–iron alloy	Sintered cobalt–iron alloy
Valve lift (mm)	10.9	10.9
Valve timing: open	12° btdc	48° bbdc
close	52° abdc	16° atdc

The compact combustion chamber is as cast, and the piston is flat topped; there are two separate squish or quench areas. The spark plug is located very close to the centre of the combustion chamber.

To ensure predictable and reliable clamping of the cylinder head to the block, a 'joint control' system is used, in which carefully toleranced bolts are loaded to their yield point. The cylinder head gasket has minimal compression relaxation, and there are stainless steel bore flanges for the cylinder bore sealing.

10.3.3 The cylinder block and associated components

The cylinder block is cast iron, and the weight is kept to a minimum (35 kg), by having a nominal wall thickness of 4.5 mm and a skirt depth of only 3 mm below the crankshaft centre-line. There are five main bearings, with bearing caps of the same material — this enables bore circularity to be maintained without resorting to a honing operation. All the shell bearings are either a lead–aluminium alloy or tri-metal on a steel backing strip.

The crankshaft is of nodular cast iron, with main bearing journal diameters of 60 mm, and crankpin or big-end journal diameters of 50 mm. All the crank-shaft journals have under-cut radiused fillets; these are deep rolled to give a fatigue strength improvement of 35 per cent.

The connecting-rod and cap are made from forged steel, with a centre distance of 151 mm. A squirt hole in the connecting-rod provides cylinder lubrication from the big-end.

The piston design is the result of finite element modelling and high-speed engine tests. The piston is made from aluminium alloy, and has cast-in-steel struts at the gudgeon pin bosses to control the thermal expansion. The top ring is made from nodular cast iron with a molybdenum-filled, radiused face for wear and scuff resistance. In contrast the second ring has a phosphate-coated, tapered face; both rings are 1.5 mm thick. The oil control ring is of a three-piece construction with a stainless steel expander, and chromium-faced side rails.

10.3.4 Combustion control

Combined electronic control is used for the ignition timing and the air/fuel mixture preparation. Signals from seven transducers are processed electronically:

  (i)   ambient air temperature
 (ii)   engine load
(iii)   carburettor throttle plate, open or closed
 (iv) engine speed
 (v)   engine coolant temperature
 (vi) exhaust gas oxygen level
(vii) engine starting.

The air fuel mixture is prepared in a twin choke carburettor with progressive choke operation. The carburettor has an electric choke, and an electronic feedback system for mixture strength control. The electronic engine-management system ensures optimum fuel-economy performance, and driveability, while still meeting stringent exhaust emission regulations.

The induction and exhaust passages were optimised by air flow testing with plastic models. The compact combustion chamber, in an aluminium alloy cylinder head, with significant squish or quench areas permits a compression ratio of 8.9:1 to be used with 91 RON fuel; the final performance is shown in figure 10.13. The engines in current production have a compression ratio of 9.6:1, or 9:1 if naturally aspirated, while the turbocharged engine has the compression ratio reduced to 8.1:1 (see appendix C, section C.7).

In all versions of the engine, emission control is by a combination of exhaust gas recirculation (EGR), air injection (exhaust manifold outlet cold; catalytic converter hot), and a catalytic converter below the exhaust manifold. The catalytic converter uses a three-way catalyst on a monolithic substrate.

10.3.5 Catalyst systems

The catalyst systems described here are produced by Johnson Matthey, a firm that has been responsible for much manufacture and development work. Currently the strictest emission controls are enforced in the USA and Japan, and the parallel development of legislation and the corresponding solutions are shown in table 10.3.

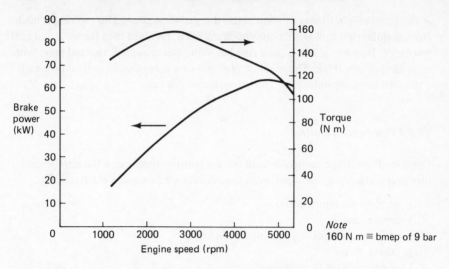

Figure 10.13   Chrysler 2.2 litre engine output (reprinted with permission, © 1981 Society of Automotive Engineers, Inc.)

Table 10.3   US Federal Emissions Limits (grams of pollutant per mile)

Model year	CO	HC	NO_x	Solution
1966	87	8.8	3.6	Pre-control
1970	34	4.1	4.0	Retarded ignition, thermal reactors, exhaust gas
1974	28	3.0	3.1	recirculation (EGR)
1975	15	1.5	3.1	Oxidation catalyst
1977	15	1.5	2.0	Oxidation catalyst and improved EGR
1980	7	0.41	2.0	Improved oxidation catalysts and three-way catalysts
1981	7	0.41	1.0	Improved three-way catalyst and support materials

The test is a simulation of urban driving from a cold start in heavy traffic. Vehicles are driven on a chassis dynamometer (rolling road), and the exhaust products are analysed using a constant-volume sampling (CVS) technique in which the exhaust is collected in plastic bags. The gas is then analysed for carbon monoxide (CO), unburnt hydrocarbons (HC) and nitrogen oxides (NO_x), using standard procedures. In 1970, three events — the passing of the American Clean Air Act, the introduction of lead-free petrol, and the adoption of cold start test cycles for engine emissions — led to the development of catalyst systems.

Catalysts usually work under carefully controlled steady-state conditions, but this is obviously not the case for engines — especially after a cold start. While catalyst systems were being developed, engine emissions were controlled by retarding the ignition and using exhaust gas recirculation (both to control NO_x) and a thermal reactor to complete oxidation of the fuel. These methods of NO_x control led to poor fuel economy and poor driveability (that is, poor transient engine response). Furthermore, the methods used to reduce NO_x emissions tend to increase CO and HC emissions and vice versa — see figure 3.14. The use of EGR and retarding the ignition also reduce the power output and fuel economy of engines.

Catalysts were able to overcome these disadvantages and meet the 1975 emissions requirements. The operating regimes of the different catalyst systems are shown in figure 10.14. With rich-mixture running, the catalyst promotes the reduction of NO_x by reactions involving HC and CO:

$$4HC + 10NO \rightarrow 4CO_2 + 2H_2O + 5N_2$$

and

$$2CO + 2NO \rightarrow 2CO_2 + N_2$$

Since there is insufficient oxygen for complete combustion, some HC and CO will remain. With lean-mixture conditions the catalyst promotes the complete oxidation of HC and CO:

$$4HC + 5O_2 \rightarrow 4CO_2 + 2H_2O$$
$$2CO + O_2 \rightarrow 2CO_2$$

With the excess oxygen, any NO_x present would not be reduced.

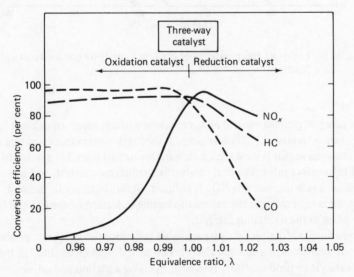

Figure 10.14   Conversion efficiencies of catalyst systems (courtesy of
                Johnson Matthey)

Oxidation catalyst systems were the first to be introduced, but NO_x emissions still had to be controlled by exhaust gas recirculation. Excess oxygen was added to the exhaust (by an air pump), to ensure that the catalyst could always oxidise the CO and HC. The requirements of the catalyst system were:

(1) High conversion of CO and HC at low operating temperatures.
(2) Durability – performance to be maintained after 80 000 km (50 000 miles).
(3) A low light-off temperature.

Light-off temperature is demonstrated by figure 10.15. The light-off temperature of platinum catalysts is reduced by adding rhodium, which is said to be a 'promoter'.

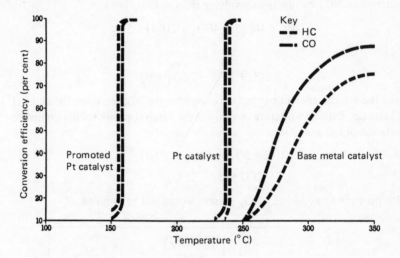

Figure 10.15   Light-off temperatures of different catalysts (courtesy of Johnson Matthey)

Dual catalyst systems control NO_x emissions without resort to exhaust gas recirculation or retarded ignition timings. A feedback system incorporating an exhaust oxygen sensor is used with a carburettor or fuel injection system to control the air/fuel ratio. The first catalyst is a reduction catalyst, and by maintaining a rich mixture the NO_x is reduced. Before entering the second catalyst, air is injected into the exhaust to enable oxidation of the CO and HC to take place in the oxidation catalyst.

Conventional reduction catalysts are liable to reduce the NO_x, but produce significant quantities of ammonia (NH_3). This would then be oxidised in the second catalyst to produce NO_x. However, by using a platinum/rhodium system the selectivity of the reduction catalyst is improved, and a negligible quantity of ammonia is produced.

Three-way catalyst systems control CO, HC and NO_x emissions as a result of developments to the platinum/rhodium catalysts. As shown by figure 10.14 very close control is needed on the air/fuel ratio. This is normally achieved by electronic fuel injection, with a lambda sensor to provide feedback by measuring the oxygen concentration in the exhaust. A typical air/fuel ratio perturbation for such a system is ±0.25 (or ±0.02ϕ).

Two types of catalyst support are used:

(1) honeycomb supports, either ceramic or metal monoliths coated with alumina
(2) alumina pellets.

Alumina pellets are obviously a simple solution, but are heavier and have a greater flow restriction. To be effective, the platinum particles have to be finely dispersed, and stabilisers are necessary to prevent aggregation of the particles.

European emission regulations are now becoming more stringent, but they are unlikely to become stricter than the USA regulations. Unlike the USA, a complete absence of leaded petrol is improbable in Europe; this leads to a significant problem for catalyst systems. Lead-tolerant catalysts are discussed by Diwell and Harrison (1981), and they also accurately forecast the current trends in European legislation. The lead-tolerant catalysts developed by Johnson Matthey are described by Anon. (1984), along with their relevance to European legislation.

10.4 Ford 2.5 litre DI diesel engine

10.4.1 Background

Compression ignition engines with direct fuel injection (DI) have typically a 10–15 per cent fuel economy benefit over indirect injection engines and are easier to start. However, indirect injection engines are normally used in light automotive applications, because the restricted speed range of direct injection engines (say up to 3000 rpm) leads to a poor power-to-weight ratio, and necessitates a multi-ratio gearbox. Indirect injection engines achieve their greater speed range by injecting fuel into a pre-chamber where there is rapid air motion. The air motion is generated in a throat that connects the pre-chamber and the main chamber; but this also leads to pressure drops and the high heat transfer coefficients that account for the reduced efficiency.

The Ford high-speed direct injection engine achieves a high engine speed (4000 rpm) by meticulous attention to the fuel injection and air motion. This four-cylinder engine is naturally aspirated, slightly over-square, with a bore of 93.7 mm and a stroke of 90.5 mm. Since there is no need to pump air into and out of the pre-chamber the compression ratio can be lower, while still maintaining good starting performance. Consequently, the compression ratio is 19:1 while the compression ratio for an indirect injection engine is typically 22:1.

10.4.2 Description

An isometric sectioned drawing of the engine is shown in figure 10.16, and the complete engine specification is given in appendix C, section C.9. A single toothed belt drives the camshaft and fuel injection pump, while V-belt drives are used for the other engine auxiliaries. The inlet and exhaust valves are all in-line, but are offset slightly from the cylinder bore axes; the induction and exhaust manifolds are on opposite sides of the engine to provide a crossflow system. The engine is inclined at an angle of $22\frac{1}{2}°$ to the vertical, and this is shown more clearly in figure 10.17, a sectioned view of the engine.

The camshaft is mounted at the side of the cylinder block and the cam followers operate the valves through short push rods and rocker arms. The valve clearance is adjusted by spherically ended screws in the rocker arms, which engage

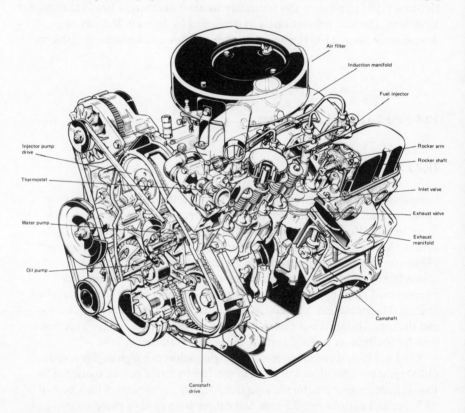

Figure 10.16   Ford 2.5 litre DI Diesel (courtesy of Ford)

with the push rods. To ensure valve train rigidity, the rocker shaft is carried by five pedestal bearings. The single valve spring rests on a hardened steel washer (to reduce wear), and the spring retainer is connected to the valve stem by a multi-groove collet. The valves operate in valve guides inserted into the cast iron cylinder head; the valve seats are hardened by induction heating after machining. The valve stem seals restrict the flow of lubricating oil down the valve stem, in order to reduce the build-up of carbon deposits.

The cylinder head for a direct injection engine is simpler than a comparable indirect injection design since there is no pre-chamber or heater plug. The cast iron cylinder head contains a dual-acting thermostat, which restricts water circulation to within the engine and heater until a temperature of 82°C is attained.

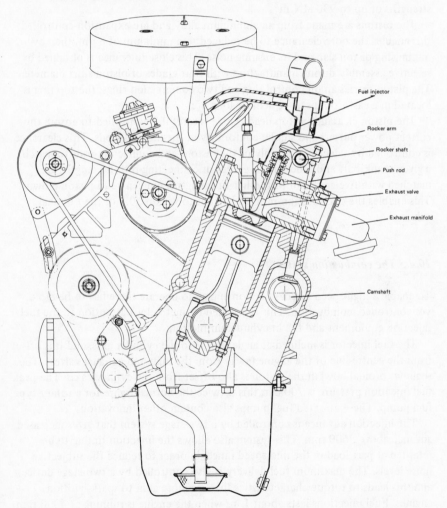

Figure 10.17   Cross-section of the Ford 2.5 litre DI Diesel (courtesy of Ford)

The water flow to the cylinder head from the cylinder block is controlled by graduated holes in the cylinder head gasket. To ensure optimum clamping of the cylinder head to the block, a torque-to-yield system is used on the cylinder head bolts.

The cylinder bores are an integral part of the cast iron cylinder block. A finite element vibration analysis was used in the design of the cylinder block to help minimise the engine noise levels. The main and big-end bearing shells use an aluminium–tin alloy. The crankshaft is made from nodular cast iron, with induction hardened journals to reduce wear. All the crankshaft journals are under-cut and fillet-rolled to improve the fatigue life. The connecting-rods are made from an air hardening steel that is relatively easy to machine, despite strengths of up to 930 MN/m^2.

The pistons are made from an aluminium alloy and are expansion-controlled; this enables the bore clearance to be reduced to a minimum of $8 \mu\text{m}$, thereby minimising piston slap and its ensuing noise. This close tolerance is obtained by selective assembly during manufacture with four grades of piston skirt diameter. The piston carries an oil control ring and two compression rings; the top ring is located in a cast iron insert to achieve long life.

The piston clearance at top dead centre is carefully controlled to ensure the correct power output and smoke conformity at the higher speeds. This clearance is controlled by selective assembly of the pistons and connecting-rods to match a given crankshaft and block. Four grades of connecting-rod (by length) are matched with five grades of piston height (gudgeon pin centre to piston crown). This enables the compression ratio to be kept to within ±0.7 of the nominal value.

10.4.3 The combustion system

To enable a high-speed direct injection engine to operate there has to be rapid, yet controlled, combustion. This is attained by meticulous attention to the fuel injection equipment and the in-cylinder air motion.

The fuel injector is inclined at angle of 23° to the cylinder axis, and it is offset from the centre-line of the engine (away from the inlet and exhaust valves). The shallow toroidal bowl in the piston is also offset from the cylinder axis. The peak fuel injection pressure is 700 bar; this is twice the normal figure for a rotary type fuel pump. These up-rated fuel pumps also contain other innovations.

The injection advance is controlled by a two-stage system that gives increased advance above 3600 rpm. This system also allows the injection timing to be retarded at part load in the mid-speed range, in order to reduce the subjective noise levels. The maximum fuel delivery is also controlled by a two-stage device, and this leads to torque characteristics that are more akin to spark ignition engines. Fuel injections lasts about 1 ms when the engine is running at 4000 rpm.

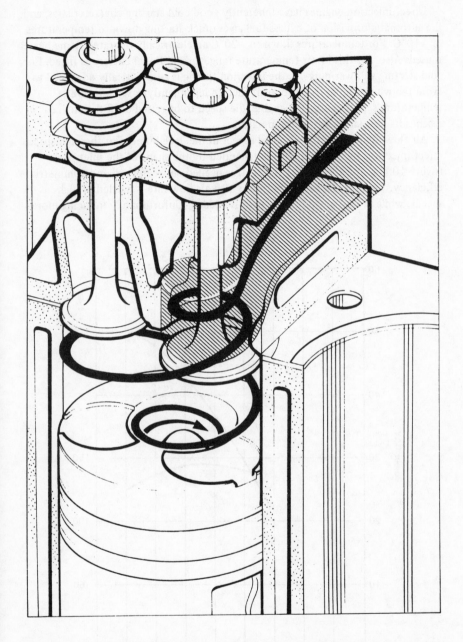

Figure 10.18   Swirl generation in the Ford 2.5 litre DI Diesel (courtesy of Ford)

Direct injection engines have inherently good cold starting characteristics, and the automatic provision of excess fuel gives quick starting down to temperatures of −10°C. For temperatures down to −20°C an electrically operated flame heater is used. Also, for this lower temperature range an additional battery is fitted. For cold starting and warming-up, the injection timing is automatically advanced to avoid the white smoke from unburnt fuel. Independently, a separate water-temperature sensor causes the engine idle speed to be raised until the engine reaches its normal operating temperature.

Air flow management is a crucial aspect in the development of the high-speed direct injection engine. The swirl is generated by the shape of the inlet port (figure 10.18), and the correct trade-off is needed between swirl and volumetric efficiency. To limit the smoke output a high air flow rate is needed at high speeds, while at low speeds a high swirl is needed. Unfortunately inducing more

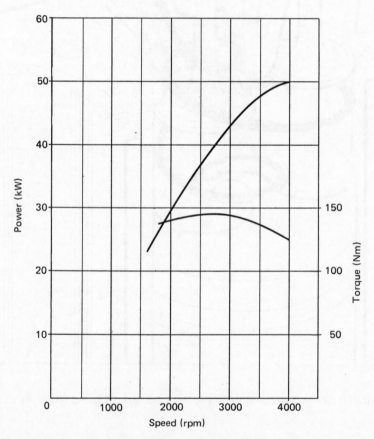

Figure 10.19   Power output of the Ford 2.5 litre DI Diesel (courtesy of Ford)

swirl reduces the volumetric efficiency. During production every cylinder head is checked for the correct air flow performance.

The aluminium induction manifold is the result of a CAD study, and the manufacture is tightly controlled to ensure accurate alignment with the ports in the cylinder head. Any misalignment between the passages in the inlet manifold and the cylinder head has a disproportionately serious effect on the flow efficiency. The consequences would be particularly serious on a direct injection engine. To improve the air flow at the valve throat, the valve seat is angled at 30°, and the valve throat is angled at 45°. During development a decrease in valve diameter was found to improve the air flow — this is explained by the reduced interference between the inlet air flow and the cylinder wall. The performance of the engine is shown in figure 10.19.

Appendix A: The Use of SI Units

SI (Système International) Units are widely used, and adopt prefixes in multiple powers of one-thousand to establish the size ranges. Using the watt (W) as an example of a base unit:

picowatt	(pW)	10^{-12}	W
nanowatt	(nW)	10^{-9}	W
microwatt	(μW)	10^{-6}	W
milliwatt	(mW)	10^{-3}	W
watt	(W)	1	W
kilowatt	(kW)	10^{3}	W
megawatt	(MW)	10^{6}	W
gigawatt	(GW)	10^{9}	W
terawatt	(TW)	10^{12}	W

It is unusual for any single unit to have such a size range, nor are the prefixes nano (10^{-9}) and giga (10^{9}) very commonly used.

An exception to the prefix rule is the base unit for mass — the kilogram. Quantities of 1000 kg and over commonly use the tonne (t) as the base unit (1 tonne (t) = 1000 kg).

Sometimes a size range using the preferred prefixes is inconvenient. A notable example is volume; here there is a difference of 10^{9} between mm^{3} and m^{3}. Consequently it is very convenient to make use of additional metric units:

$$1 \text{ cm} = 10^{-2} \text{ m}$$

thus

$$1 \text{ cm}^{3} = 10^{3} \text{ mm}^{3} = 10^{-6} \text{ m}^{3}$$

$$1 \text{ litre (l)} = 1000 \text{ cm}^{3} = 10^{6} \text{ mm}^{3} = 10^{-3} \text{ m}^{3}$$

Pressure in SI units is the unit of force per unit area (N/m^{2}), and this is some-times denoted by the Pascal (Pa). A widely used unit is the bar (1 bar = 10^{5} N/m^{2}), since this is nearly equal to the standard atmosphere:

$$1 \text{ standard atmosphere (atm)} = 1.10325 \text{ bar}$$

A unit commonly used for low pressures is the torr:

$$1 \text{ torr} = \frac{1}{760} \text{ atm}$$

In an earlier metric system (cgs), 1 torr = 1 mm Hg.

The unit for thermodynamic temperature (T) is the kelvin with the symbol K (*not* °K). Through long established habit a truncated thermodynamic temperature is used, called the Celsius temperature (t). This is defined by

$$t = (T - 273.15) \,°\text{C}$$

Note that (strictly) temperature differences should always be expressed in terms of kelvins.

Some additional metric (non-SI) units include:

Length	1 micron = 10^{-6} m
	1 angstrom (Å) = 10^{-10} m
Force	1 dyne (dyn) = 10^{-5} N
Energy	1 erg = 10^{-7} N m = 10^{-7} J
	1 calorie (cal) = 4.1868 J
Dynamic viscosity	1 poise (P) = 1 g/cm s = 0.1 N s/m^2
Kinematic viscosity	1 stokes (St) = 1 cm^2/s = 10^{-4} m^2/s

A very thorough and complete set of definitions for SI Units, with conversions to other unit systems, is given by Haywood (1972).

Conversion factors for non-SI units

Exact definitions of some basic units:

Length	1 yard (yd) = 0.9144 m
Mass	1 pound (lb) = 0.453 592 37 kg

$$\text{Force} \qquad 1 \text{ pound force (lbf)} = \frac{9.806\,65}{0.3048} \text{ pdl}$$

(1 poundal (pdl) = 1 lb ft/s^2)

Most of the following conversions are approximations:

Length	1 inch (in) = 25.4 mm
	1 foot (ft) = 0.3048 m
	1 mile (mile) $\approx$ 1.61 km
Area	1 square inch (sq. in) = 645.16 mm^2
	1 square foot (sq. ft) $\approx$ 0.0929 m^2

Volume 1 cubic inch (cu. in) $\approx$ 16.39 cm^3
1 gallon (gal) $\approx$ 4.546 l
1 US gallon $\approx$ 3.785 l

Mass 1 ounce (oz) $\approx$ 28.35 g
1 pound (lb) $\approx$ 0.4536 kg
1 ton (ton) $\approx$ 1016 kg
1 US short ton $\approx$ 907 kg

Density 1 lb/ft^3 $\approx$ 16.02 kg/m^3

Force 1 pound force (lbf) $\approx$ 4.45 N

Pressure 1 lbf/in^2 $\approx$ 6.895 kN/m^2
1 in Hg $\approx$ 3.39 kN/m^2
1 in H$_2$O $\approx$ 0.249 kN/m^2

Dynamic viscosity 1 lb/ft s $\approx$ 1.488 kg/m s
N s/m^2

Kinematic viscosity 1 ft^2/s $\approx$ 0.0929 m^2/s

Energy 1 ft lbf $\approx$ 1.356 J

Power 1 horse power (hp) $\approx$ 745.7 W

Specific fuel
consumption 1 lb/hp h $\approx$ 0.608 kg/kW
$\approx$ 0.169 kg/MJ

Torque 1 ft lbf $\approx$ 1.356 N m

Energy 1 therm (= 10^5 Btu) $\approx$ 105.5 MJ

Temperature 1 rankine (R) = $\dfrac{1}{1.8}$ K

$$\left\{ \begin{array}{l} t_F = (T_R - 459.67)^\circ F \\ \text{thus } t_F + 40 = 1.8\,(t_C + 40) \end{array} \right\}$$

Specific heat capacity
Specific entropy $\Big\}$ 1 Btu/lb R = 4.1868 kJ/kg K

Specific energy 1 Btu/lb = 2.326 kJ/kg

Appendix B: Answers to Numerical Problems

2.1 0.678, 0.623, 0.543, 0.405

2.3 6.0 bar, 33.5 per cent, 22.3:1

2.4 6.9 bar, 20.7 per cent, 73 per cent

2.5 (a) 74.1 kW, 173.6 N m; (b) 9.89 bar, 7.27 bar, 10.99 bar, 8.08 bar; (c) 25.25 per cent, 58.3 per cent; (d) 90.7 per cent, 12.0:1

3.2 C: 0.855, H: 0.145; 13.50:1

3.3 C: 0.848, H: 0.152; 0.833; 15:1

3.4 Insufficient data are given to answer the question accurately, so clearly stated assumptions (and their significance) are more important than the numerical values.
702 K, 21.3 bar; 4500 K, 137 bar

3.5 (i) CO: 2.15, O_2: 1.08, CO_2: 6.77 bar; (ii) CO: 4.12, O_2: 2.06, CO_2: 3.81 bar

3.6 51.1

6.1 29.4 per cent

7.1 54°C, 671°C, 92 per cent

7.4 $$\left[\left(\frac{p_2}{p_1}\right)^{(\gamma_a-1)/\gamma_a} - 1\right] = \left[1 - \left(\frac{p_4}{p_3}\right)^{(\gamma_e-1)/\gamma_e}\right] \eta_{mech}\, \eta_c\, \eta_t \left(1 + \frac{1}{AFR}\right)\left(\frac{c_{p_e}}{c_{p_a}}\right)\left(\frac{T_3}{T_1}\right)$$

Appendix C: Engine Specifications

C.1 Sulzer marine compression ignition engines

Two-stroke engine: Sulzer RTA series (Wolf (1982))

bore diameter	380	to	840	mm
speed range	190	to	70	rpm
power range	1.72	to	35.52	MW

For example, Sulzer RTA 84

bore diameter	840 mm	
rating	R1	R4
cylinder mep	15.35	15.53 bar
peak cylinder pressure	125	125 bar
specific fuel consumption	173	167 g/kWh
power	2960	2100 kW/cylinder
speed	87	70 rpm

Four-stroke engine: Sulzer ZA40 (figure C.1) (Lustgarten (1982))

	Maximum continuous rating	Economy rating
output	640	600 kW/cylinder
nominal speed	560	560 rpm
mean piston speed	8.96	8.96 m/s
bmep	22.74	21.32 bar
specific fuel consumption ⎫ peak cylinder pressure ⎬ ⎭	figure C.2	
bore	400 mm	
stroke	480 mm	

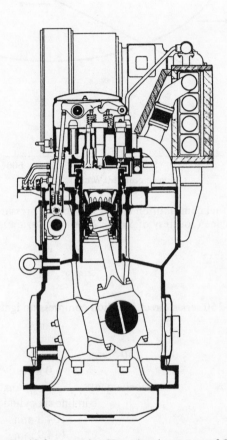

Figure C.1   Sulzer ZA40 four-stroke CI engine (courtesy of *Sulzer Technical Review*)

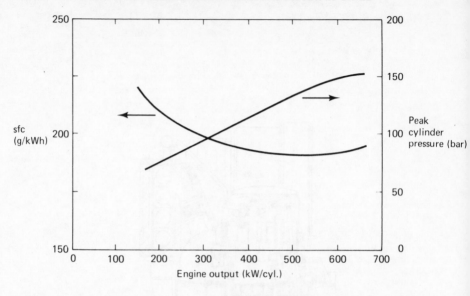

Figure C.2   Specific fuel consumption and peak cylinder pressure of the Sulzer
ZA40 engine (courtesy of *Sulzer Technical Review*)

C.2 Ford 'Dover' 90–150 series direct injection compression ignition engines

	150 Series	*110 Series*
nominal capacity		6.0 litres
engine type	turbocharged	naturally aspirated
engine configuration		in-line six-cylinder
bore		104.8 mm
stroke		114.9 mm
swept volume		5.947 litres
compression ratio	15.5:1	15.9:1
power Gross BS Au 141a	114; 2400	85; 2600 kW; rpm
DIN 70020	110; 2400	81; 2600 kW; rpm
torque Gross BS Au 141a	495; 1700	240; 1500 N m; rpm
DIN 70020	480; 1700	335; 1500 N m; rpm
cooling system capacity		
standard/heavy-duty	27.2/29.7	24.85/26.35 litres
oil system capacity	15.6	15.9/15.3 litres
injection order		1–5–3–6–2–4
weight (dry)	488	467 kg

C.3 Rolls Royce CV12 direct injection compression ignition engine

This engine is described by Hartley (1976).

engine configuration	V12
displacement	26.1 litres
size	1.48 × 1.02 × 1.37 m
weight	1860 kg
power output	
automotive (2100 rpm)	350–615 kW
industrial (1800 rpm)	400 kW
marine (1800 rpm)	470 kW
military (2300 rpm)	about 895 kW
peak pressures (commercial engines)	125 bar
specific fuel consumption	
full load	225 g/kWh
maximum economy	207 g/kWh

Peak pressures in the military engine are controlled by reducing the compression ratio from 14:1 to 12:1. Since the lower compression ratio can lead to starting difficulties and poor running at light load, the air can be pre-heated. When the air manifold temperature falls below 200°C, a CAV device is used that incorporates a fuel injector and high-energy igniter in the manifold. A pressure ratio of about 3:1 is used on the military engine.

C.4 Ford V6 'Essex' spark ignition engine

engine configuration	V6
bore	93.7 mm
stroke	72.4 mm
swept volume	2.994 litres
compression ratio	8.9:1
brake power (max.)	100 kW at 5500 rpm
torque (max.)	233 Nm at 3000 rpm
firing order	1(R)-4(L)-2(R)-5(L)-3(R)-6(L)
	(No. 1 cylinder is furthest from the fly-wheel; (R) – right, (L) – left, as viewed from the flywheel)
static ignition timing	10° btdc
valve timing	
inlet opens	29° btdc

inlet closes	67° abdc
exhaust opens	70° btdc
exhaust closes	14° atdc
valve lift	
inlet	9.35 mm
exhaust	8.61 mm
valve clearances (hot)	
inlet	0.25 mm
exhaust	0.45 mm
lubrication system	
pump type	eccentric bi-rotor or sliding vane
oil filter	external full flow, pressure-relief type
oil pressure	3.1–3.4 bar (gauge)
cooling system	
pressurised water	0.7 bar (gauge)

C.5 Chrysler 875 cm^3 spark ignition engine

	standard	sport
engine configuration	in-line four-cylinder	
bore	68 mm	
stroke	60.2 mm	
swept volume	0.875 litres	
compression ratio	10:1	
brake power (max.)	29; 5000	35; 6100 kW; rpm
torque (max.)	70.5; 2800	70.5; 4300 Nm; rpm
firing order	1–3–4–2	
static ignition timing	4° btdc	
valve timing		
inlet opens	6° btdc	23° btdc
inlet closes	46° abdc	53° abdc
exhaust opens	46° bbdc	61° bbdc
exhaust closes	6° atdc	15° atdc
valve size		
inlet	30.5	32.4 mm
exhaust	27.0	27.0 mm
carburation	single solex	twin stromberg
exhaust system	branched manifold	tuned system
lubrication system		
pump type	lobed rotor	
oil filter	external full flow	
cooling system	pressurised water	

The combustion chamber and cylinder head arrangement is very similar to that shown in figure 6.2. The engine derives from Coventry Climax work in the 1940s, and thus has resemblances to the valve train design of Jaguars. Comparison between the standard and sport engine models shows that the extra power is obtained by improving the torque at high engine speeds. This is achieved by maintaining high volumetric efficiency at high speeds, through the use of more sophisticated induction and exhaust systems, a slightly larger inlet valve, and a camshaft with greater valve overlap.

C.6 Jaguar V12 spark ignition engine

	V12	V12 HE
engine configuration	V12	V12 HE
bore	90 mm	
stroke	70 mm	
swept volume	5.345 litres	
compression ratio	9:1 (7.8:1 USA)	12.5:1 (11.5:1 USA)
brake power (max.)	203; 5850	223; 5500 kW; rpm
torque (max.)	412; 3600	432; 3000 Nm; rpm
fuel system	electronic fuel injection	
ignition system	constant-energy electronic ignition	
lubrication system	crankshaft-driven crescent pump	
	relief flow through oil-cooler	

C.7 Chrysler 2.2 litre spark ignition engine

	EDE	EDJ	EDF	EDG
engine configuration	in-line four-cylinder			
bore	87.5 mm			
stroke	92.0 mm			
swept volume	2.20 litres			
type	naturally aspirated			turbocharged
compression ratio	9.0:1	9.6:1	9.0:1	8.1:1
brake power (max.)	72; 5200	82; 5600	74; 5600	106; 5600 kW; rpm
torque (max.)	161; 3200	175; 3600	164; 3200	217; 3200 N m; rpm
firing order	1-3-4-2			
ignition system	electronic ignition			
fuel system	carburettor			electronic fuel injection
lubrication system	3.45 bar (gauge) at 2000 rpm			
oil capacity	3.8 litres			
coolant system	pressurised water, 0.96-1.24 bar (gauge)			
coolant capacity	8.2 litres			

C.8 Fiat–Sofim indirect injection compression ignition engine

Sofim is a partnership of Fiat, Alfa-Romeo and Saviem. The 8130, 8140 and 8160 engines are three-cylinder, four-cylinder and six-cylinder engines, respectively, using common engine components; the engines are designed for light automotive, marine and industrial applications – see figure C.3. The engines are described by Torazza (1979), but only the four-cylinder unit is presented here.

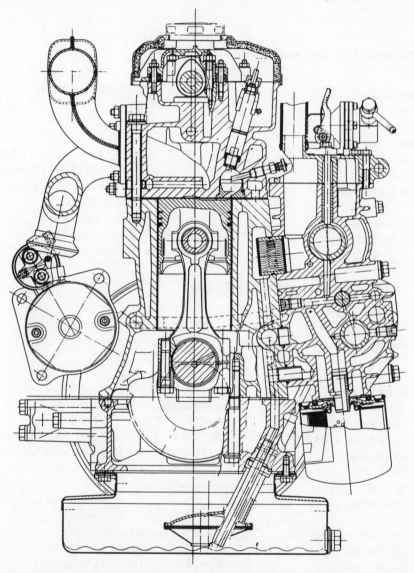

Figure C.3   Fiat–Sofim indirect injection engine (from Torazza (1979))

8140 naturally aspirated   *8140 turbocharged*

engine configuration	in-line four-cylinder	
bore	93 mm	
stroke	90 mm	
swept volume	2.444 litres	
compression ratio	22:1	
brake power (max.)	51.5 kW at 4200 rpm	also	62.5 kW at 4200 rpm
torque (max.)	148 N m at 2500 rpm	see	192 N m at 2500 rpm
sfc (min.)	258 g/kWh	figure C.4	258 g/kWh
combustion system	Ricardo Comet	
fuel injection system	Bosch distributor type pump	
turbocharger	Garret 7992 T3 with waste gate	
lubrication system	4 bar (gauge)	

C.9 Ford 2.5 litre DI diesel engine

engine configuration	in-line four-cylinder
bore	93.7 mm
stroke	90.5 mm
swept volume	2.496 litres
compression ratio	19:1
brake power (max.) {see also	50; 4000 kW; rpm
torque (max.) { figure 10.19	143; 2700 N m; rpm
valve periods	
inlet	232°
exhaust	244°
lubrication system capacity	7.0 litres
cooling system	
pump type	centrifugal impeller
system pressure	0.5 bar

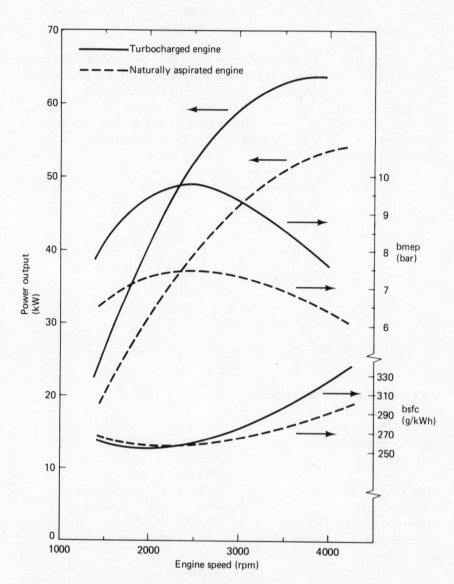

Figure C.4 Performance of the Fiat–Sofim 2.445 indirect ignition engine
(adapted from Torazza (1979))

C.10 Volvo–Penta 2-stroke spark ignition outboard motor engines

	Volvo Penta 400	*Volvo Penta 700*
engine configuration	in-line three-cylinder	
bore	60	80 mm
stroke	58.8	60 mm
swept volume	0.50	0.90 litres
brake power (max.)	30; 6000	50; 5500 kW; rpm
for racing applications	65; 9000	— kW; rpm
engine weight (dry)	24.2	45 kg
sfc		
wide open throttle	375	402 g/kWh
part load	349	375 g/kWh
ignition system	capacitive breakerless type	

Data from Stillerud (1978).

Appendix D: Stratified Charge Engines

Stratified charge engines were introduced in chapter 1, section 1.4.2, but the following discussion is presented here because it draws on material developed in chapters 2, 3, 4 and 5.

The search for both better fuel consumption and lower emissions than those of the spark ignition engine led to the development of many types of stratified charge engine. In a conventional spark ignition engine the air/fuel mixture is assumed to be homogeneous, and only a fairly limited range of mixture strengths can be ignited. Consequently, the output at any speed is controlled by throttling, a process that reduces the flow rates of both air and fuel. Unfortunately, the pressure drop across the throttle (the 'throttling loss') increases the work needed to draw in the charge ('pumping work'), and this reduces the engine efficiency. Furthermore, if power output could be regulated solely by reducing the quantity of fuel, the leaner air/fuel mixture would have a higher corresponding cycle efficiency (chapter 2, section 2.5). With regard to emissions, the weaker mixtures should reduce the emissions of carbon monoxide (CO) and oxides of nitrogen (NO_x), but hydrocarbon emissions (HC) will still depend on the effectiveness of the combustion.

Conventional spark ignition engines will not ignite mixtures weaker than an air/fuel ratio of about 17:1, high compression ratio lean-burn engines will ignite mixtures with an air/fuel ratio of 25:1, but stratified charge engines can ignite mixtures with an air/fuel ratio weaker than 50:1. Combustion in stratified charge engines is achieved by having a close-to-stoichiometric mixture around the spark plug, and a much weaker mixture in the major part of the combustion chamber, a mixture that would not normally be ignited by a spark. Combustion starts in the richer mixture, and is then sufficiently vigorous to propagate into the remaining lean mixture.

There are two alternative approaches to producing a stratified charge:

(1) a single combustion chamber with in-cylinder fuel injection
(2) a divided combustion chamber.

With fuel injection the combustion is controlled by the rate and timing of fuel injection, the air motion, and the spark timing. Fuel injection can be used in conjunction with a carburetted weak mixture, or with injection into air alone.

Examples of single-chamber stratified charge engines have been developed by many manufacturers. Figure D.1 shows the Texaco Controlled Combustion System (TCCS) and the MAN FM system. As with compression ignition engines, careful matching of the fuel and air mixing is essential.

There are various examples of divided combustion chamber stratified charge engines; some are shown in figure D.2. The Honda CVCC (compound vortex controlled combustion) is the best known type, since it is the only engine to have entered full production. In the Honda engine a third valve controls the supply of a rich carburetted mixture to the pre-chamber, while the main inlet valve controls the supply of a weak mixture to the main part of the combustion chamber. An alternative system is to use fuel injection into the pre-chamber, and admit either air or a weak carburetted mixture to the main chamber. A computer model for these types of combustion systems is described by Wall and Heywood (1978).

A comprehensive survey of engine emissions and fuel economy is given by Blackmore and Thomas (1977). While NO_x and CO emissions are intrinsically low, unburnt hydrocarbons are a more serious problem, although these can be minimised by an oxidation catalyst in the exhaust system.

The stratified charge engine has failed to become widely used for several reasons:

(1) The hoped for economy and emissions levels have not been achieved; this may be through lack of sufficient development work.
(2) The power output is reduced; this occurs by definition with weak mixtures or where the charge is not homogeneous — as for compression ignition engines.

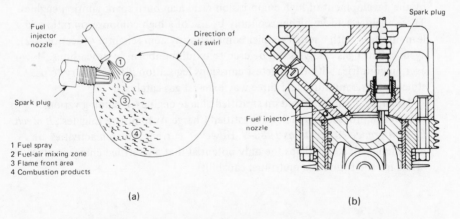

1 Fuel spray
2 Fuel-air mixing zone
3 Flame front area
4 Combustion products

(a)   (b)

Figure D.1 Single-chamber stratified charge engines. (a) Texaco controlled combustion system TCSS (plan view) (from Campbell (1978)); (b) MAN FM system (with acknowledgement to Newton *et al.* (1983))

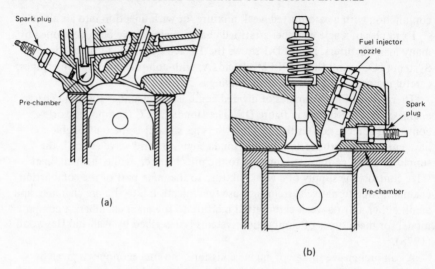

Figure D.2 Pre-chamber stratified charge engines. (a) Honda CVCC (compound vortex controlled combustion); (b) Mercedes Benz (from Campbell (1978))

(3) The added cost of fuel injection equipment or additional carburation equipment, and in some cases pre-chambers, makes the construction as expensive as compression ignition engines.

In some ways the development of stratified charge engines has been overtaken by the development of high compression ratio lean-burn spark ignition engines. These engines achieve higher economy by use of a high compression ratio and lean mixtures with a much simpler combustion system (chapter 4, section 4.2); again exhaust gas catalysts can be used to oxidise unburnt hydrocarbons. However, in countries with the strictest emissions legislation it is necessary to use stoichiometric mixtures and three-way exhaust gas catalysts.

Much has been published on stratified charge engines, including various conference proceedings such as 'Stratified Charge Automotive Engines', *I. Mech. E., Conference Publications 1980–9*. However, current research activities are much reduced, and perhaps the only potential that a stratified charge engine offers for the future is a multi-fuel capability.

Appendix E: Engine Tuning

E.1 Introduction

Engine tuning can be aimed at improving economy or improving power output; the two aims do not have to be mutually exclusive, but usually are, in the quest for ultimate power output. Most of the techniques follow quite logically from material covered in the main part of this text.

The output and fuel economy of compression ignition engines can both be increased significantly by turbocharging, but this is a specialised field (discussed in chapter 7). Apart from this, tuning is more a question of adequate maintenance; the main points are:

(1) the correct spray pattern and injector operating pressure (chapter 5, section 5.5.1)
(2) correct timing and calibration of the fuel pump (chapter 5, section 5.5.2)
(3) adequate compression (that is, insignificant leakage past the valves and piston rings)
(4) unblocked air filter.

The following remarks are directed primarily at spark ignition engines, since these have the greatest scope for tuning.

E.2 Tuning for spark ignition engine economy

Normally tuning is limited to maintenance. The substitution or addition of carburettors only benefits economy when the new carburettor is correctly matched and the original was poorly matched. The addition of electronic ignition will often assist starting, and will certainly help to overcome the common (but very significant) problem of worn and badly adjusted contact breakers.

The effect of vehicle maintenance on fuel economy is treated at some length by Blackmore and Thomas (1977). Some of the key items (not necessarily in order of importance) that affect spark ignition engine fuel economy are:

(i)   idling mixture setting
(ii)   correct idling speed
(iii)   static ignition timing
(iv)   dwell angle (controlled by the contact breaker gap)
(v)   ignition timing, vacuum advance
(vi)   ignition timing, centrifugal advance
(vii)   spark plug condition
(viii) thermostat operation
(ix)   air cleaner condition
(x)   valve gear adjustment

The relative significance of these items will vary from one application to the next, and also with the usage – the correct idling mixture and speed are particularly significant in low-speed, stop/start urban driving. Many of these items can now be controlled by electronic engine-management systems. These not only ensure optimum initial fuel economy with minimal deterioration, but they also control engine emissions – often the justification for their use.

Engine maladjustment is surprisingly common; results are quoted from Blackmore and Thomas (1977) in table E.1.

Table E.1   Incidence of parameters outside manufacturer's specification on a sample of 72 vehicles

Correction needed	Percentage of sample
Mixture strength at idle	83.4
Static ignition timing	75.0
Dwell angle (contact breaker gap)	40.6
Valve gear adjustment	29.2
Spark plug replacement	23.6
Contact breaker replacement	20.8
Mixture strength at 2000 rpm	18.1
Cylinder leakage	16.7
Air cleaner replacement	5.6

Ignition timing measurements are fairly straightforward, and mixture strength is readily determined by measuring the carbon monoxide (CO) content in the exhaust. If carbon monoxide measuring equipment is not available, a Colourtune spark plug is very useful. This spark plug has a glass insulator so that the combustion can be seen. A rich mixture is characterised by a yellow flame from glowing carbon particles, while a weak mixture is characterised by the blue flame that is associated with the oxidation of carbon monoxide to carbon dioxide. Colourtune spark plugs are also useful for checking inter-cylinder variation in mixture strength.

The averaged reduction in fuel economy from controlled malfunctions on a range of engines is shown in table E.2.

Table E.2 Effect of engine maladjustment on fuel economy

Factor	Percentage reduction in fuel economy
One failed spark plug	12.6
Mixture strength from weak to rich	10.5
Idling speed increased from 650 to 850 rpm	5.1
Seized centrifugal advance mechanism	13.4
Failed vacuum advance device	2.8
Removal of thermostat during warm-up	2.1
Restricted air cleaner element	11.5

Finally, it must be remembered that correct vehicle maintenance is necessary for optimum performance: the brakes must not be allowed to rub, and the tyres must be correctly inflated. The widespread change from cross-ply to radial tyres has improved road-holding, and has also increased fuel economy by as much as 10 per cent.

E.3 Tuning for spark ignition engine output

After a discussion of 'blueprinting', various engine modifications are discussed in an approximate order of cost-effectiveness. The cost and effectiveness will vary significantly from one engine to the next, and will depend on the initial state of tune, the intended application, and the availability of special parts.

'Blueprinting' is when the engine is not changed from the manufacturer's specification, but every clearance or tolerance is adjusted to optimise the performance. The crankshaft and cylinder block are chosen to give the largest swept volume. The crankshaft is checked for straightness and balance, and the optimum bearing clearances will be chosen. The connecting-rods are reduced to the minimum weight and correct weight distribution, and are also checked for any bend or twist. The clearance volumes in each combustion chamber are checked and equalised to provide the same (high) compression ratio. It is common practice to polish the combustion chamber and gas passages — this may not reduce the fluid frictional losses, but it certainly inhibits the build-up of combustion deposits.

The induction and exhaust passages are checked for smoothness, particularly at junctions. An exaggerated example that would cause a significant reduction in volumetric efficiency is shown in figure E.1, along with a correct example.

The following aspects of tuning all involve changes to engine components.

Multi-branch exhaust manifolds are discussed in chapter 6, section 6.4, and these improve the gas exchange process; free-flow exhaust systems can also be fitted.

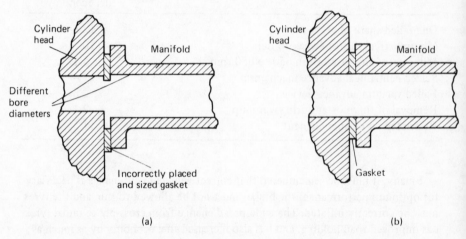

Figure E.1   Manifold and cylinder head assembly. (a) Badly fitted assembly;
(b) correctly fitted assembly

Multi-carburettor installations improve the volumetric efficiency of the engine, and can help to ensure better mixture distribution — see chapter 6, section 6.4. Care is needed to ensure that the flows of fuel and air through each carburettor are the same. The length of the induction pipe before the carburettor can also be tuned.

Camshafts are often available in several forms, with the more highly tuned camshafts providing the greatest valve overlap and lift — see chapter 6, section 6.2. The power output is gained by increasing the torque at high speeds. The maximum torque may well be reduced, but the speed at which it occurs will be increased. These changes have an adverse effect on low speed and part load operation because of increased charge dilution by exhaust gas residuals. High lift camshafts are also likely to increase the wear in the valve train. In some cylinder heads it may also be possible to increase the size of the valves.

Compression ratio increases are usually limited unless special fuels are available that are less susceptible to self-ignition (knock) — see chapter 3, section 3.5. The compression ratio is raised by either fitting special pistons, fitting longer connecting-rods, or machining material from the mating face of the cylinder head (or block).

In all cases it is essential to obtain the correct mixture and distribution as well as the optimum ignition timing. In practice, the ignition timing may well be further advanced than the manufacturer's specification. A manufacturer has to ensure that an engine subject to manufacturing tolerances will have knock-free operation throughout its range. In certain applications it may be acceptable for some knock to occur at full throttle and low speeds, if this is part of the operating range that is not likely to be used.

Turbocharging is a very effective but specialised method of increasing the engine output — it is the subject of chapter 7. However, turbocharging kits are available for many applications and in addition there are firms that provide a specialist service. The problems with turbochargers are: matching the disparate flow characteristics of the engine and turbocharger; ensuring that the higher pressures and temperatures do not lead to self-ignition of the fuel; and minimising the 'turbolag' that arises from the inertia of the turbocharger rotor. It is almost inevitable, especially with higher boost pressures, that the compression ratio is lowered in order to limit the cycle pressures and temperatures; this is despite the adverse effect this has on fuel economy. The incoming charge can also be cooled by:

(1) Inter-coolers.
(2) Special fuels such as methanol that have a high enthalpy of evaporation (and higher octane rating). The cooling effect of methanol is about twice that of petrol.
(3) Injection of other fluids — notably water, with its very high enthalpy of evaporation, and nitrous oxide (N_2O), which dissociates and cools the charge and, since it also provides additional oxygen, enables a richer fuel mixture to be burnt.

In racing applications, very high boost pressures are used, and charge cooling is essential. An example is the BMW 1.5 litre Forced Induction F1 engine:

	boost pressure	*power output*
race conditions	2.9 bar	475 kW
qualifying conditions	3.2 bar	520 kW

However, it is quite acceptable for this type of engine to be rebuilt after each race.

Bibliography

The most prolific source of published material on internal combustion engines is the Society of Automotive Engineers (SAE) of America. Some of the individual papers are selected for inclusion in the annual *SAE Transactions*. Other SAE publications include the *Progress in Technology* (PT) and *Specialist Publications* (SP), in which appropriate papers are grouped together. Examples are

> SP-532 *Aspects of Internal Combustion Engine Design*
> PT-24 *Passenger Car Diesels*

The SAE also organise a wide range of meetings and conferences, and publish the magazine *Automotive Engineering*.

In the United Kingdom the Institution of Mechanical Engineers (I. Mech. E.) publish *Proceedings* and hold conferences, some of which relate to internal combustion engines. The Automobile Division also publishes the bi-monthly *Automotive Engineer*.

The other main organisers of European conferences include:

CIMAC   Conseil International des Machines à Combustion
FISITA Fédération International des Sociétés d'Ingénieur et de Techniciens de l'Automobile
IAVD   International Association for Vehicle Design
ISATA   International Symposium on Automotive Technology and Automation

Many books are published on internal combustion engines, and this can be seen in the list of references. However, since books can become dated, care and discretion are necessary in the use of old material. Two books worth looking out for are mentioned below.

(1) Professor J. B. Heywood is preparing a book on internal combustion engines. This can be considered as a successor to *The Internal Combustion Engine in Theory and Practice* (Taylor, 1966, 1968).
(2) Volume I of *The Thermodynamics and Gas Dynamics of Internal Combustion Engines* by R. S. Benson (edited by J. H. Horlock and D. E.

Winterbone; OUP, 1982) has been published, but Volume II is not yet available. Volume I provides a thorough and comprehensive treatment of processes external to the engine cylinder, and numerical solutions are described along with FORTRAN listings.

Finally, *Engines – the search for power* by John Day (published by Hamlyn, London, 1980) is a copiously illustrated book describing the development of all types of engine.

References

Annual Book of ASTM Standards, Part 47 — Test Methods for Rating Motor, Diesel and Aviation Fuels, ASTM, Philadelphia, Pennsylvania

L. F. Adams (1975), *Engineering Measurements and Instrumentation*, EUP, London

A. Allard (1982), *Turbocharging and Supercharging*, Patrick Stephens, Cambridge

W. J. D. Annand (1963), 'Heat Transfer in the Cylinders of Reciprocating Internal Combustion Engines', *Proc. I. Mech. E.*, Vol. 177, No. 36, pp. 973–90

W. J. D. Annand and G. E. Roe (1974), *Gas Flow in the Internal Combustion Engine*, Foulis, Yeovil

Anon. (1984), 'Catalytic exhaust-purification for Europe?', *Automotive Engineer*, Vol. 9 No. 1

BS2637: 1978 Motor and aviation-type fuels — Determination of knock characteristics — Motor method, B.S.I., London

BS2638: 1978 Motor fuels — Determination of knock characteristics — Research method, B.S.I., London

BS2869: 1970 Petroleum fuels for oil engines and burners, B.S.I., London

BS4040: 1978 Petrol (gasoline) for motor vehicles, B.S.I., London

A. Baker (1979), *The Component Contribution*, Hutchinson Benham, London

R. S. Benson and N. D. Whitehouse (1979), *Internal Combustion Engines*, Pergamon, Oxford

D. R. Blackmore and A. Thomas (1977), *Fuel Economy of the Gasoline Engine*, Macmillan, London

C. Campbell (1978), *The Sports Car*, 4th edn, Chapman and Hall, London

D. F. Caris and E. E. Nelson (1958), 'A new look at high compression engines', *SAE* paper No. 61A

H. Cohen, G. F. C. Rogers and H. I. H. Saravanamuttoo (1972), *Gas Turbine Theory*, 2nd edn, Longman, London

T. Crisp (1984), 'Jaguar V12 HE engine', private communication

H. Daneshyar (1976), *One-Dimensional Compressible Flow*, Pergamon, Oxford

G. O. Davies (1983), 'The Preparation and Combustion Characteristics of Coal Derived Transport Fuels', Paper C85/83, *Int. Conf. on Combustion in Engineering*, Vol. II, MEP, London

A. F. Diwell and B. Harrison (1981), 'Car Exhaust Catalysts for Europe', *Platinum Metals Review*, Vol. 25, No. 4

M. J. Donnelly, J. Junday and D. H. Tidmarsh (1981), 'Computerised data acquisition and processing system for engine test beds', *3rd Int. Conf. on Automotive Electronics*, MEP, London

D. Downs and R. W. Wheeler (1951–52), 'Recent developments in knock research', *Proc. I. Mech. E. (AD)*, Pt III, p. 89

D. Downs, S. T. Griffiths and R. W. Wheeler (1961), 'The part played by the preparational stage in determining lead anti-knock effectiveness', *J. Inst. Petrol.*, Vol. 47, p. 1

Ford (1982), *Ford Energy Report*, Interscience Enterprises, Channel Islands, UK

R. J. Francis and P. N. Woollacott (1981), 'Prospects for improved fuel economy and fuel flexibility in road vehicles', *Energy Paper No. 45*, Department of Energy, HMSO, London

A. G. Gaydon and H. G. Wolfhard (1979), *Flames, their Structure, Radiation and Temperature*, 4th edn, Chapman and Hall, London

P. E. Gliken, D. F. Mowbray and P. Howes (1979), 'Some Developments on Fuel Injection Equipment for Diesel Engine Powered Cars', *I. Mech. E. Conf. Publications 1979-13*, MEP, London

E. M. Goodger (1979), *Combustion Calculations*, Macmillan, London

A. B. Greene and G. G. Lucas (1969), *The Testing of Internal Combustion Engines*, EUP, London

D. A. Greenhalgh (1983), 'Gas phase temperature and concentration diagnostics with lasers', *Int. Conf. on Combustion in Engineering*, Vol. I, I. Mech. E. Conf. Publications 1983-3, MEP, London

R. T. C. Harman (1981), *Gas Turbine Engineering*, Macmillan, London

J. Hartley (1976), 'On the warpath with powerful compact diesels', *The Engineer*, 3 June

R. W. Haywood (1972), *Thermodynamic Tables in SI (Metric) Units*, 2nd edn, CUP, Cambridge

R. W. Haywood (1980), *Analysis of Engineering Cycles*, 3rd edn, Pergamon International Library, Oxford

J. B. Heywood (1980), 'Engine Combustion Modelling – An Overview', in J. N. Mattavi and C. A. Amann (Eds), *Combustion Modelling in Reciprocating Engines*, Plenum Press, New York

J. B. Heywood, J. M. Higgins, P. A. Watts and R. J. Tabaczynski (1979), 'Development and Use of a Cycle Simulation to Predict SI Engine Efficiency and NO_x Emissions', *SAE 790291*

M. H. Howarth (1966), *The Design of High Speed Diesel Engines*, Constable, London

A. P. Ives and M. V. Trenne (1981), 'Closed loop electronic control of diesel fuel injection', *3rd Int. Conf. on Automotive Electronics*, I. Mech. E., Conf. Publications 1981-10, MEP, London

A. W. Judge (1967), *High Speed Diesel Engines*, 6th edn, Chapman and Hall, London

A. W. Judge (1970), *Motor Manuals 2: Carburettors and Fuel Injection Systems*, 8th edn, Chapman and Hall, London

P. S. Katsoulakos (1983), 'Effectiveness of the combustion of emulsified fuels in diesel engines', *Int. Conf. on Combustion in Engineering*, Vol. II, I. Mech. E. Conf. Publications 1983-3, MEP, London

B. E. Knight (1960–61), 'Fuel injection system calculations', *Proc. I. Mech. E.*, No. 1

B. Lewis and G. von Elbe (1961), *Combustion Flames and Explosions of Gases*, 2nd edn, Academic Press, New York

G. A. Lustgarten (1982), 'The Sulzer ZA40 Engine, a Further Development of the Well-proven Z40', *Sulzer Technical Review*, Vol. 1

J. N. Mattavi and S. A. Amann (1980), 'Combustion Modelling in Reciprocating Engines', Plenum Press, New York

M. May (1979), 'The High Compression Lean Burn Spark Ignited 4-stroke Engine', *I. Mech. E. Conf. Publications 1979-9*, MEP, London

H. Mundy (1972), 'Jaguar V12 Engine: Its Design and Development History', *Proc. I. Mech. E.*, Vol. 186, paper 34/72, pp. 463-77

K. Newton, W. Steeds and T. K. Garrett (1983), *The Motor Vehicle*, 10th edn, Butterworths, London

J. P. Packer, F. J. Wallace, D. Adler and E. R. Karimi (1983), 'Diesel fuel jet mixing under high swirl conditions', *Int. Conf. on Combustion in Engineering*, Vol. II, I. Mech. E. Conf. Publications 1983-3, MEP, London

D. A. Parker and M. Kendrick (1974), 'A camshaft with variable lift–rotation characteristics', Paper B-1-11, *15th FISITA Congress*, Paris

K. Radermacher (1982), 'The BMW Eta Engine Concept', *Proc. I. Mech. E.*, Vol. 196

H. R. Ricardo and J. G. G. Hempson (1968), *The High Speed Internal Combustion Engine*, 5th edn, Blackie and Son, London

G. F. C. Rogers and Y. R. Mayhew (1980a), *Engineering Thermodynamics, Work and Heat Transfer*, 3rd edn, Longman, London

G. F. C. Rogers and Y. R. Mayhew (1980b), *Thermodynamic and Transport Properties of Fluids, SI Units*, 3rd edn, Blackwell, Oxford

P. H. Smith (1967), *Valve Mechanisms for High Speed Engines*, Foulis, Yeovil

P. H. Smith (1968), *Scientific Design of Exhaust and Intake Systems*, Foulis, Yeovil

K. G. H. Stillerud (1978), 'Aspects of two-stroke design viewpoints from outboard developments', *Design and Development of Small Internal Combustion Engines*, I. Mech. E., Conf. Publication 1978-5, MEP, London

R. J. Tabaczynski (1983), 'Turbulence measurements and modelling in reciprocating engines – an overview', *Int. Conf. on Combustion in Engineering*, Vol. I, I. Mech. E. Conf. Publications 1983-3, MEP, London

C. F. Taylor (1966), *The Internal Combustion Engine in Theory and Practice*, Vol. I, M.I.T. Press

C. F. Taylor (1968), *The Internal Combustion Engine in Theory and Practice*, Vol. II, M.I.T. Press

G. Torazza (1979), 'Fiat–Sofim engines' evolution in the early eighties', *The Passenger Car Power Plant of the Future*, I. Mech. E. Conf. Publications 1979-13, MEP, London

J. C. Wall and J. B. Heywood (1978), 'The Influence of Operating Variables and Prechamber Size on Combustion in a Prechamber Stratified-Charge Engine', *SAE 780966*

F. J. Wallace, M. Tarabad and D. Howard (1983), 'The differential compound engine – a new integrated engine transmission system for heavy vehicles', *Proc. I. Mech. E.*, Vol. 197A

N. Watson and M. S. Janota (1982), *Turbocharging the Internal Combustion Engine*, Macmillan, London

N. Watson, S. Wijeyakumar and G. L. Roberts (1981), 'A microprocessor controlled test facility for transient vehicle engine system development', *3rd Int. Conf. on Automotive Electronics*, MEP, London

W. L. Weertman and J. W. Dean (1981), 'Chrysler Corporation's New 2.2 Liter 4 Cylinder Engine', *SAE 810007*

J. A. Whitehouse and J. A. Metcalfe (1956), 'The influence of lubricating oil on the power output and fuel consumption of modern petrol and compression ignition engines', MIRA Report No. 2

G. Wolf (1982), 'The Large Bore Diesel Engine', *Sulzer Technical Review*, Vol. 3

G. Woschni (1967), 'A Universally Applicable Equation for the Instantaneous Heat Transfer Coefficient in the Internal Combustion Engine', *SAE Transactions*, Vol. 76, p. 3065, paper 670931

Index